Lectures on

TOPICS IN STOCHASTIC DIFFERENTIAL EQUATIONS

By

DANIEL W. STROOCK

Lectures delivered at the

INDIAN INSTITUTE OF SCIENCE, BANGALORE

under the

T.I.F.R.—I.I.SC. PROGRAMME IN APPLICATIONS OF MATHEMATICS

Notes by

SATYAJIT KARMAKAR

Published for the

TATA INSTITUTE OF FUNDAMENTAL RESEARCH, BOMBAY

SPRINGER-VERLAG

Berlin Heidelberg New York

1982

Author

DANIEL W. STROOCK
Department of Mathematics
University of Colorado
BOULDER, COLORADO 80309
U.S.A.

ISBN 3-540-11549-8 Springer-Verlag, Berlin. Heidelberg. New York
ISBN 0-387-11549-8 Springer-Verlag, New York. Heidelberg. Berlin

Printed by N. S. Ray at The Book Centre Limited, Sion East, Bombay 400 022 and published by H. Goetze, Springer-Verlag, Heidelberg, West Germany

PRINTED IN INDIA

P R E F A C E

THESE NOTES are based on five weeks of lectures given during December 1980 and January 1981 at T.I.F.R. in Bangalore. My purpose in these lectures was to provide some insight into the properties of solutions to stochastic differential equations. In order to read these notes, one need only know the basic Itô theory of stochastic integrals. For example, H.P. McKean's Stochastic Integrals (Academic Press, 1969) contains all the background material.

After developing a little technical machinery, I have devoted Chapter I to the study of solutions of S.D.E.'s as a function of the initial point. This topic has been discussed by several authors; the treatment given here is based on a recent paper by H. Kunita (to appear in the Proceedings of the 1980 L.M.S. conference at Durham). In fact, the only way in which the present treatment differs from this is that I have been a little more careful about the integrability estimates.

Chapter II is devoted to the study of solutions as a function of time. A large part of this material is adapted from my work with S.R.S. Varadhan ("On the support of diffusion processes with applications to the strong maximum principle", Proc. 6th Berkeley Symp. on Math. Stat. and Prob., Vol.III (1970)). The presentation of this material has been greatly aided by the incorporation of ideas introduced by Y. Takahashi and S. Watanabe in their paper "The probability functionals of diffusion processes" (to appear in the Proceedings of the 1980 L.M.S. conference at Durham). To introduce the reader to their work, in the last section of Chapter II , I have derived a very special case of the general result derived in the paper by Takahashi and Watanabe.

I would like to thank S. Karmakar for his efforts in guiding these notes to completion. Also, I am grateful to R. Karandikar for his useful comments on these notes. Finally, it gives me pleasure to express my gratitude to K.G. Ramanathan and the Tata Institute for the opportunity to visit India.

CONTENTS

CHAPTER I

SOLUTIONS TO STOCHASTIC DIFFERENTIAL EQUATIONS AS A FUNCTION OF THE STARTING POINT

§1. BASIC INEQUALITIES OF MARTINGALE THEORY:

The most important inequality in the martingale theory is the Doob's inequality which states that if $(X(t), F_t, P)$ is a right continuous, integrable sub-martingale, then

(1.1) $$P(\sup_{0\leq t\leq T} X(t) \geq \lambda) < \frac{1}{\lambda} E[X(T), \sup_{0\leq t\leq T} X(T) \geq \lambda], \ \lambda > 0,$$

for all $T > 0$.

As a consequence of (1.1), one has, assuming that $X(.) \geq 0$:

(1.2) $$E[\sup_{0\leq t\leq T} X(t)^p]^{1/p} \leq \frac{p}{p-1} E[X(T)^p]^{1/p}, \ 1 < p < \infty .$$

The passage from (1.1) to (1.2) goes as follows :

Let $X(T)^* = \sup_{0\leq t\leq T} X(t)$. Then using (1.1) we have

$$\begin{aligned} E[(X(T)^*)^p] &= p\int_0^\infty \lambda^{p-1} P(X(T)^* \geq \lambda)d\lambda \\ &\leq p\int_0^\infty \lambda^{p-2} E[X(T), X(T)^* \geq \lambda]d\lambda \\ &= p\int_0^\infty \lambda^{p-2} d\lambda \int_0^\infty P(X(T) \geq \mu, X(T)^* \geq \lambda)d\mu. \\ &= \frac{p}{p-1} \int_0^\infty E[X(T)^*)^{p-1}, X(T) \geq \mu]d\mu . \\ &= \frac{p}{p-1} E[X(T)^*)^{p-1} X(T)] \\ &\leq \frac{p}{p-1} E[(X(T)^*)^p]^{1-1/p} E[X(T))^p]^{1/p} \end{aligned}$$

where we have used Hölder's inequality to obtain the last line. Now dividing both sides by $E[(X(T)^*)^p]^{1-1/p}$, we get the required result.

The basic source of continuous martingales is stochastic integrals. Let (β_t, F_t, P) be a 1-dimensional Brownian motion and $\theta(.)$ be a $F.$-progressively measurable $\mathbb{R}$-valued function satisfying

$$E[\int_0^T |\theta(t)|^2 dt] < \infty$$

for all $T > 0$. Set

$$X(t) = \int_0^t \theta(s) d\beta(s).$$

Then $(X(t), F_t, P)$ and $(X^2(t) - \int_0^t \theta(s)^2 ds, F_t, P)$ are continuous martingales. In particular,

$$E[X^2(T)] = E[\int_0^T \theta(t)^2 dt] \quad \text{for all } T > 0.$$

By (1.2), this means that

$$E[(X(T)^*)]^{1/2} \leq 2E[\int_0^T \theta(t)^2 dt]^{1/2}.$$

The following theorem contains an important generalization of this observation.

(1.3) <u>THEOREM</u> (*Burkholder*): *Let* $(\beta(t), F_t, P)$ *be a 1-dimensional Brownian motion and* $\sigma(.)$ *be a* $F.$*-progressively measurable* $R^n \times R^d$ *-valued function satisfying*

$$E[\int_0^T \text{Trace } a(t) dt] < \infty, T > 0$$

where $a(.) = \sigma(.)\sigma(.)^*$. *Set* $X(t) = \int_0^t \sigma(s) d\beta(s)$. *Then for* $2 \leq p < \infty$ *and* $T > 0$,

$$E[\sup_{0\le t\le T} |X(t)|^p]^{1/p} < C_p E[\int_0^T \text{Trace } a(t)dt)^{p/2}]^{1/p}$$

where

$$C_p = \left(\frac{p^{p+1}}{2(p-1)^{p-1}}\right)^{1/2}$$

Proof. By Ito's formula:

$$|X(t)|^p = p\int_0^t |X(s)|^{p-2} \sigma(s)\, X(s)\, d\beta(s)$$
$$+ \frac{1}{2}\int_0^t p|X(s)|^{p-2} (\text{Trace } a(s)+(p-2) \frac{<X(s),a(s)X(s)>}{|X(s)|^2} ds$$

Thus,

$$E[|X(T)|^p] = \frac{1}{2} E[\int_0^T p|X(t)|^{p-2} (\text{Trace } a(t)+(p-2) \frac{<X(t),a(t)X(t)>}{|X(t)|^2} dt]$$
$$\le \frac{p(p-1)}{2} E[\int_0^T |X(t)|^{p-2} \text{ Trace } a(t)\, dt]$$
$$\le \frac{p(p-1)}{2} E[\sup_{0\le t\le T} |X(t)|^{p-2} \int_0^T \text{Trace } a(t)dt]$$
$$\le \frac{p(p-1)}{2} E[\sup_{0\le t\le T} |X(t)|^p]^{1-2/p} E[(\int_0^T \text{Trace } a(t))^{p/2}]^{2/p}$$

and so

$$E[\sup_{0\le t\le T} |X(t)|^p] < (\frac{p}{p-1})^p E[|X(T)|^p]$$
$$\le \frac{p^{p+1}}{2(p-1)^{p-1}} E[\sup_{0\le t\le T} |X(t)|^p]^{1-2/p} E[(\int_0^T \text{Trace } a(t)^{p/2}]^{2/p}$$

Now dividing both sides by $E[\sup_{0\le t\le T} |X(t)|^p]^{1-2/p}$ we get the required result.

§2. SOLUTIONS TO STOCHASTIC DIFFERENTIAL EQUATION AS A FUNCTION OF (t,x):

Let $\sigma: R^n \to R^n \times R^d$ and $b: R^n \to R^n$ be measurable functions satisfying

$$\| \sigma(x) - \sigma(y) \|_{H.S.} \leq L|x-y|, \quad x,y \in R^n$$

(2.1) $$|b(x)-b(y)| \leq L|x-y|, \quad x,y \in R^n$$

where $L < \infty$ and $\| . \|_{H.S.}$ denotes the Hilbert Schmidt norm.*

Also assume that $(\beta(t), F_t(P)$ is a d-dimensional Brownian motion.

(2.2) THEOREM. *For each* $x \in R^n$ *there exists a unique solution* $\xi(.,x)$ *to*

(2.3) $$\xi(t,x) = x + \int_0^t \sigma(\xi(s,x))d\beta(s) + \int_0^t b(\xi(s,x))ds, \quad t \geq 0.$$

In fact, for each $2 \leq p < \infty$ *and* $T > 0$ *there are* $A_p(T) < \infty$ *such that*

(2.4) $$E[\sup_{0 \leq t \leq T} |\xi(t,x) - \xi(t,y)|^p] \leq A_p(T)(|x-y|^p)$$

Proof. The existence of solution is proved by the Picard iteration. Define $\xi_0(.) = x$ and

$$\xi_{n+1}(t) = x + \int_0^t \sigma(\xi_n(s))d\beta(s) + \int_0^t b(\xi_n(s)ds$$

Then for $2 \leq p < \infty$,

---

* $\|A\|_{H.S.}$ is the Hilbert-Schmidt norm of the matrix A. That is, $\|A\|^2_{H.S.} = \text{Trace } AA^*$.

$$E[\sup_{0\leq t\leq T} |\xi_{n+1}(t) - \xi_n(t)|^p]^{1/p}$$

$$\leq E[|\int_0^T b(\xi_n(t)) - b(\xi_{n-1}(t)dt|^p]^{1/p}$$

$$+ E[\sup_{0\leq t\leq T} |\int_0^T (\sigma(\xi_n(s)) - \sigma(\xi_{n-1}(s)))d\beta(s)|^p]^{1/p}$$

$$\leq L\int_0^T E[|\xi_n(t) - \xi_{n-1}(t)|^p]^{1/p}\,dt$$

$$+ C_p \mathbb{E}[(\int_0^T |\xi_n(t) - \xi_{n-1}(t)|^2 dt)^{p/2}]^{1/p}$$

$$\leq L\int_0^T E[|\xi_n(t) - \xi_{n-1}(t)|^p]^{1/p}\,dt$$

$$+ C_p\ L\ T^{p-2/2p}\ E[\int_0^T |\xi_n(t) - \xi_{n-1}(t)|^p\ dt]^{1/p}$$

and so

$$E[\sup_{0\leq t\leq T} |\xi_{n+1}(t) - \xi_n(t)|^p]$$

$$\leq 2^{p-1}\ L^p(T^{p-1} + C_p^p\ T^{p-1/2})\int_0^T E[\sup_{0\leq t\leq T} |\xi_n(s)-\xi_{n-1}(s)|^p]dt.$$

From this, it follows by induction that

$$E[\sup_{0\leq t\leq T} |\xi_{n+1}(t) - \xi_n(t)|^p] < \frac{B_p(T)}{n!}$$

where $B_p(T) < \infty$. Hence $\xi_n(.)$ converges uniformly on finite intervals and clearly the limit satisfies (2.3). In fact, we have shown that there is a solution $\xi(.)$ such that

$$E[\sup_{0\leq t\leq T} |\xi(t)|^p] < \infty$$

for all $T > 0$ and $1 \leq p < \infty$.

To prove uniqueness, note that if $\eta(.)$ is a second solution and $\tau_R = \inf\{t \geq 0 : |\eta(.) - x| \geq R\}$, then

$$E[|\eta(t \wedge \tau_R) - \xi(t \wedge \tau_R)|^2]$$

$$\leq 2L^2 E[\int_0^{t\wedge\tau_R} |\eta(s)-\xi(s)|^2 ds] + 2L^2 tE[\int_0^{t\wedge\tau_R} |\eta(s)-\xi(s)|^2 ds]$$

and so $\eta(t \wedge \tau_R) = \xi(t \wedge \tau_R)$ (a.s., P) for all t and R. Since $\eta(.)$ and $\xi(.)$ are P - a.s. continuous and $P(\sup_{0\leq t\leq T} |\xi(t)| < \infty) = 1$ for all $T > 0$, uniqueness is now obvious.

Finally,

$$E[\sup_{0\leq t\leq T} |\xi(t,y) - \xi(t,x)|^p]^{1/p}$$

$$\leq |x-y| + C_p L\, E[(\int_0^T |\xi(t,y) - \xi(t,x)|^2 dt)^{p/2}]^{1/p}$$

$$+ L \int_0^T E[|\xi(t,y) - \xi(t,x)|^p]^{1/p}\, dt$$

and so

$$E[\sup_{0\leq t\leq T} |\xi(t,y) - \xi(t,x)|^p]$$

$$\leq 3^{p-1}|x-y|^p + 3^{p-1}L^p(C_p^p\, T^{p-1/p}) \int_0^T E[\sup_{0\leq t\leq T} |\xi(s,y)-\xi(s-x)|^p]dt$$

Therefore

$$E[\sup_{0\leq t\leq T} |\xi(t,y) - \xi(t,x)|^p]$$

$$\leq 3^{p-1}\, |x-y|^p \exp(3^{p-1}\, L^p \int_0^T (C_p^p\, t^{(p-1)/2} + t^{p-1})dt).$$

This proves our theorem.

The importance of (2.4) is that it allows us to find a version of $\xi(t,x)$ which is a.s. continuous with respect to (t,x). To see how this is done, we need the following real-variable lemma.

(2.5) LEMMA : *Let* P *and* Ψ *be strictly increasing continuous function on* $[0,\infty]$ *such that* $P(0) = \Psi(0) = 0$ and $\Psi(\infty) = \infty$. *Also suppose that* L *is a normed linear space and that* $f: R^d \to L$ *is strongly continuous on* $B(a,r)$ $(\equiv \{x \in R^d: |x-a| < r\})$. *Then*

$$\int_{B(a,r)} \int_{B(a,r)} \Psi\left(\frac{\|f(x)-f(y)\|}{P(|x-y|)}\right) dx\, dy \leq B$$

implies that

$$\|f(x) - f(y)\| \leq 8 \int_0^{|x-y|} \Psi^{-1}\left(\frac{4^{d+2}}{\gamma^2} \frac{B}{u^{2d}}\right) P(du), \quad x,y \in B(a,r)$$

where

$$\gamma = \inf_{x \in B(a,r)} \inf_{1 < \rho \leq 2} \frac{|B(x,\rho) \cap B(a,1)|}{\rho^d}$$

Proof. Define

$$I(x) = \int_{B(a,r)} \left(\frac{\|f(x) - f(y)\|}{P(|x-y|)}\right) dy .$$

Given distinct points $x,y \in B(a,r)$, set $\rho = |x-y|$ and choose $c \in B(\frac{x+y}{2}, \frac{\rho}{2}) \cap B(a,r)$ so that

$$I(c) \leq 2^{d+1} \frac{B}{\gamma\rho^d} .$$

This is possible because $\int_{B(a,r)} I(c)dc \leq B$. Set $x_o = y_o = c$. Now we choose x_n and y_n for $n \geq 1$ as follows. Given x_{n-1} and y_{n-1}, define d_{n-1} and e_{n-1} by

$$P(d_{n-1}) = \frac{1}{2} P(2|x_{n-1} - x|) \text{ and } P(e_{n-1}) = \frac{1}{2} P(2|y_{n-1} - y|).$$

Now choose $x_n \in B(x, \frac{1}{2} d_{n-1}) \cap B(a,r)$ and $y_n \in B(y, \frac{1}{2} e_{n-1}) \cap B(a,r)$ so that

$$I(x_n) \leq \frac{2^{d+1} B}{\gamma\, d_{n-1}^d} \quad\text{and}\quad \Psi\left(\frac{\|f(x_n)-f(x_{n-1})\|}{P(|x_n - x_{n-1}|)}\right) \leq \frac{2^{d+1}\, I(x_{n-1})}{\gamma\, d_{n-1}^d}$$

$$I(y_n) \leq \frac{2^{d+1} B}{\gamma\, e_{n-1}^d} \quad\text{and}\quad \Psi\left(\frac{\|f(y_n) - f(y_{n-1})\|}{P(|y_n - y_{n-1}|)}\right) \leq \frac{2^{d+1}\, I(y_{n-1})}{\gamma\, e_{n-1}^d}$$

This is possible by the same reasoning as that used to find c.

Note that $P(d_n) = \frac{1}{2} P(2|x_n - x|) < \frac{1}{2} P(d_{n-1})$. Thus d_n decreases to zero. Also, for $n \geq 1$:

$$\|f(x_n)-f(x_{n-1})\| \leq \Psi^{-1}\left(\frac{4^{d+2} B}{\gamma^2 d_{n-1}^d\, d_{n-2}^d}\right) P(|x_n - x_{n-1}|)$$

$$\leq \Psi^{-1}\left(\frac{4^{d+2} B}{\gamma^2 d_{n-1}^{2d}}\right) P(|x_n - x_{n-1}|)$$

where $d_{-1} = \rho$. Since $2P(d_{n-1}) = P(2|x_{n-1} - x|)$, $d_{n-1} < 2|x_{n-1} - x|$. Thus

$$P(|x_n - x_{n-1}|) \leq P(2|x_{n-1} - x|) = 2P(d_{n-1})$$

$$= 4(P(d_{n-1}) - \tfrac{1}{2}P(d_{n-1}))$$

$$\leq 4(P(d_{n-1}) - P(d_n)).$$

We therefore have:

$$\|f(x_n)-f(x_{n-1})\| \leq 4\Psi^{-1}\left(\frac{4^{d+2} B}{\gamma^2 d_{n-1}^{2d}}\right)(P(d_{n-1})-P(d_n))$$

$$\leq 4 \int_{d_n}^{d_{n-1}} \Psi^{-1}\left(\frac{4^{d+2} B}{\gamma^2\, u^{2d}}\right) P(du)$$

and so

$$\|f(x) - f(c)\| \leq 4 \int_0^\rho \Psi^{-1}\left(\frac{4^{d+2} B}{\gamma^2 u^{2d}}\right) P(du).$$

The same argument yields

$$\|f(x) - f(c)\| \leq 4 \int_0^\rho \Psi^{-1}\left(\frac{4^{d+2} B}{\gamma^2 u^{2d}}\right) P(du)$$

This proves our lemma.

(2.6) LEMMA : *Let* $X(x)$, $x \in R^d$, *be a family of Banach space-valued random variables with the properties that for some* $\alpha > 0$ *and* $p \geq d+\alpha$

(2.7) $$E[\|X(x) - X(y)\|^p] \leq C|x-y|^{d+\alpha}, \quad x,y \in R^d.$$

Then there is a family $\tilde{X}(x)$, $x \in R^d$ *such that* $\tilde{X}(x) = X(x)$ a.s. *for each* $x \in R^d$ *and* $x \to \tilde{X}(x)$ *is* a.s. *strongly continuous.*

Proof. For each $N \geq 0$, let $X^{(N)}(.)$ denote the multilinear extension of the restriction of $X(.)$ to the lattice $\{k/2^N : k \in \mathbb{Z}^d\}$. Then it is not hard to check that there is a C' (depending on C, p and α) such that

(2.8) $$E[\|X^{(N)}(x) - X^{(N)}(y)\|^p] \leq C'|x-y|^{d+\alpha}$$

Now let $\rho = (2d + \alpha/2)/p$. Then, by (2.8) :

$$\sup_N E\left[\int_{B(0,r)} \int_{B(0,r)} \left(\frac{\|X^{(N)}(x) - X^{(N)}(y)\|}{|x-y|^\rho}\right)^p dx\, dy\right] \leq C'' r^{2d}$$

where $C'' < \infty$ depends on C' and d. By (2.5), this means that for $L > 0$

$$\sup_N P\left(\|X^{(N)}(x) - X^{(N)}(y)\| \leq KL^{1/p}|x-y|^{\alpha/2p},\ x,y \in B(0,r)\right) \geq 1 - \frac{C'' r^{2d}}{L},$$

where K depends on d,p and α. Notice that

$$\sup_{\substack{x,y \in B(0,R) \\ x \neq y}} \frac{\|X^{(N)}(x) - X^{(N)}(y)\|}{|x-y|^{\alpha/2p}}$$

is a non-decreasing function of N. Hence

$$P\left\{\sup_N \sup_{\substack{x,y \in B(0,R) \\ x \neq y}} \frac{\|X^{(N)}(x) - X^{(N)}(y)\|}{|x-y|^{\alpha/2p}} \leq KL^{1/p}\right\} \geq 1 - \frac{C'' r^{2d}}{L}.$$

Since $X^{(n)}\left(\frac{k}{2^n}\right) = X\left(\frac{k}{2^n}\right)$ for all $N \geq 0$ and $k \in \mathbb{Z}^d$, $X^{(N)}(.)$ converges a.s., uniformly on $B(0,r)$ to a continuous function $\tilde{X}(.)$ which coincides with $X(x)$ at $x = k/2^N \in B(0,r)$. Since

$$E[\|X(x) - X(y)\|^p] \to 0 \quad \text{as} \quad |x-y| \to 0,$$

it is clear that $\tilde{X}(x) = X(x)$ a.s. for all $x \in B(0,r)$. Finally, since r was arbitrary, the proof is complete.

(2.9) <u>Exercise</u>: Let $f(x) = x \vee a$. Then $f'(x) = \chi_{[a,\infty]}(x)$ and $f''(x) = \delta_a(x)$ (Dirac's δ-function). Thus by "Itô's formula", if (β_t, F_t, P) is a one-dimensional Brownian motion:

$$\frac{1}{2}\int_0^t \delta_a(\beta(s))ds = \beta(t) \vee a - \int_0^t \chi_{[a,\infty]}(\beta(s))d\beta(s)$$

That is, we would suspect that

$$\lim_{\varepsilon \downarrow 0} \frac{1}{4\varepsilon}\int_0^t \chi_{[a-\varepsilon,\ a+\varepsilon]}(\beta(s))ds = \beta(t) \vee a - \int_0^t \chi_{[a,\infty]}(\beta(s))d\beta(s).$$

To check this, define

$$\ell_a(t) = \beta(t)\vee a - \int_0^t \chi_{[a,\infty]}(\beta(s))d\beta(s).$$

By the technique which we have been using, show that there is a version $L_a(t)$ of $\ell_a(t)$ which is a.s. continuous in $(t,x) \in [0,\infty] \times R$. Check that for $f \in C_0^\infty(R)$

$$\int_R f(a)L_a(t)da = F(\beta(t)) - \int_0^t F(\beta(s))d\beta(s)$$

where $F(x) = \int (x\vee a)\ f(a)da$.

From this conclude that

$$\int_R f(a)L_a(t)da = \frac{1}{2}\int_0^t f(\beta(s))ds$$

for all such f's. The identification of $L_a(t)$ as

$$\lim_{\varepsilon \downarrow 0} \frac{1}{4\varepsilon} \int \chi_{[-\varepsilon+a,\ \varepsilon+a]}(\beta(s))ds$$

is now easy.

(2.10) Exercise: Again let (β_t, F_t, P) be a 1-dimensional Brownian motion. Given $t > 0$ and $\varepsilon > 0$, let $N_\varepsilon(t)$ be the number of times that $|\beta(.)|$ goes from ε to 0 during $[0,t]$. (That is, $N_\varepsilon(t) \geq n$ if and only if there exist $0 < u_1 < v_1 < \ldots < u_n \leq v_n \leq t$ such that $|\beta(u_j)| = \varepsilon$ and $|\beta(v_j)| = 0$ for $1 \leq j \leq n$). Show that

$$N_\varepsilon(t) \to 2L_0(t) \quad \text{a.s.} \quad \varepsilon \downarrow 0.$$

The idea is the following: First show that

$$|\beta(t)| = \int_0^t \operatorname{sgn}\ \beta(s)\ d\beta(s) + 2L_0(t).$$

Next define $\tau_0 = 0$ and for $n \geq 1$

$$\sigma_n = \inf \{t \geq \tau_{n-1} : |\beta(t)| = \varepsilon\}$$

$$\tau_n = \inf \{t \geq \sigma_n : |\beta(t)| = 0\} .$$

Then

$$\sum_1^\infty \left(|\beta(\tau_n \wedge t)| - |\beta(\sigma_n \wedge t)|\right) = -\varepsilon N_\varepsilon(t) + (|\beta(t)| - \varepsilon) \sum_1^\infty \chi_{[\sigma_n, \tau_n]}(t).$$

At the same time:

$$\sum_1^\infty \left(|\beta(\tau_n \wedge t)| - |\beta(\sigma_n \wedge t)|\right) = |\beta(t)| - \sum_1^\infty \int_{\tau_{n-1}\wedge t}^{\sigma_n \wedge t} \operatorname{sgn} \beta(s) d\beta(s) - 2L_0(t) .$$

Thus

$$\varepsilon N_\varepsilon(t) - 2L_0(t) = -|\beta(t)| \sum_1^\infty \chi_{[\tau_{n-1}, \sigma_n]}(t) - \sum_1^\infty \chi_{[\sigma_n, \tau_n]}(t)$$

$$+ \sum_1^\infty \int_{\tau_{n-1}\wedge t}^{\sigma_n \wedge t} \operatorname{sgn} \beta(s)\, d\beta(s).$$

From this check that

$$E[|\varepsilon N_\varepsilon(t) - 2L_0(t)|^2] \leq c_t\, \varepsilon,\ 0 < \varepsilon \leq 1;$$

and therefore that

$$\frac{1}{n^2} N_{1/n^2}(t) \to 2L_0(t) \text{ a.s. as } n \to \infty.$$

Finally, note that if $1/(n+1)^2 \leq \varepsilon < 1/n^2$, then

$$\frac{1}{(n+1)^2} N_{1/n^2}(t) \leq \varepsilon N_\varepsilon(t) \leq 1/n^2\, N_{1/(n+1)^2}(t),$$

and so $\varepsilon N_\varepsilon(t) \to 2L_0(t)$ a.s. as $\varepsilon \downarrow 0$.

(2.11) <u>LEMMA</u>: *Let* $(\beta(t), F_t, P)$ *be a* d*-dimensional Brownian motion and suppose that* $\sigma(.)$ *and* $b(.)$ *are as in* (2.1). *Then there is a choice of* $\xi(t,x)$ *solving* (2.3) *such that*

$$(t,x) \to \xi(t,x)$$

is a.s. *continuous.*

From now on, $\xi(t,x)$ refers to the map in (2.11). We next want to discuss the mapping $x \to \xi(t,x)$ for fixed $t > 0$. We will first show that a.s. the maps $x \to \xi(t,x)$ are 1-1 and continuous for all $t \geqslant 0$.

(2.12) <u>LEMMA</u>: *Let* $T > 0$ *and* p R *be given. Then there is a* $C_p(T) < \infty$ *such that*

(2.13) $E[|\xi(t,x) - \xi(t,y)|^p] \leqslant C_p(T)\ |x-y|^p,\ t \in [0,T]$ *and* $x,y \in R^n$.

<u>Proof</u>. Set $f(z) = |z|^p$ for $z \in R^n \setminus \{0\}$. Then

$$\frac{\partial f}{\partial z_i} = p|z|^{p-2}\, z_i$$

and

$$\frac{\partial^2 f}{\partial z_i \partial z_j} = p(p-2)|z|^{p-4}\, z_i z_j + \delta_{ij}\, p|z|^{p-2}$$

Now let $x \neq y$ be given and for $0 < \varepsilon < |x-y|$ let

$$\zeta_\varepsilon = \inf\{t \geqslant 0:\ |\xi(t,x)-\xi(t,y)| \leqslant \varepsilon\}$$

and define

$$z_\varepsilon(t) = \xi(t \wedge \zeta, x) - \xi(t \wedge \zeta_\varepsilon, y).$$

Now by Itô's formula:

$$|z_\varepsilon(t)|^p - \sum_{i=1}^{n}\int_o^{t\wedge\zeta_\varepsilon}(b^i(\xi(s,x))-b^i(\xi(s,y)))\frac{\partial f}{\partial x_i}(z_\varepsilon(s))ds$$

$$- \frac{1}{2}\sum_{i,j=1}^{n}\int_o^{t\wedge\zeta_\varepsilon}(\sum_{\ell=1}^{d}\sigma_\ell^i(\xi(s,x)-\xi(s,y))\sigma_\ell^j(\xi(s,x)-\xi(s,y)))$$

$$\frac{\partial^2 f}{\partial x_i\partial x_j}(z_\varepsilon(s))\,ds$$

is a martingale. Notice that for $0 \leq s \leq \zeta_\varepsilon$:

$$|\sum_{i=1}^{n}(b^i(\xi(s,x)) - b^i(\xi(s,y)))\frac{\partial f}{\partial x_i}(z_\varepsilon(s))|$$

$$\leq |p|L|z_\varepsilon|^{p-1}\sum_{i=1}^{n}|z_{\varepsilon,i}(s)| \leq n^{1/2}L|p||z_\varepsilon(s)|^p.$$

Also for $0 \leq s \leq \zeta_\varepsilon$:

$$\sum_{i,j=1}^{n}\sum_{\ell=1}^{d}(\sigma_\ell^i(\xi(s,x))-\sigma_\ell^i(\xi(s,y)))(\sigma_\ell^j(\xi(s,x))-\sigma_\ell^j(\xi(s,y)))\frac{\partial^2 f}{\partial x_i\partial x_j}(z_\varepsilon(s))$$

$$\leq dL^2p(p-2)|z_\varepsilon(s)|^{p-2}\sum_{i,j=1}^{n}z_{\varepsilon,i}(s)z_{\varepsilon,j}(s)+dL^2p|z_\varepsilon(s)|^p$$

$$\leq (dn\,L^2p(p-2) + dL^2p)|z_\varepsilon(s)|^p .$$

Thus

$$E[|z_\varepsilon(t)|^p.] \leq |x-y|^p + A_{d,n}(p)\int_o^t E[|z_\varepsilon(s)|^p]ds,$$

and so

$$E[|\xi(t\wedge\zeta_\varepsilon,x)-\xi(t\wedge\zeta_\varepsilon,y)|^p] \leq |x-y|^p\,e^{A_{d,n}(p)\,t}$$

Letting $\varepsilon \downarrow 0$, we see that

$$E[|\xi(t\wedge\zeta_\varepsilon,x) - \xi(t\wedge\zeta_\varepsilon,y)|^p] \leq |x-y|^p\,e^{A_{d,n}(p)\,t}$$

where $\zeta = \inf \{t > 0: \xi(t,x) = \xi(t,y)\}$. Taking $p = -1$, we conclude that $P(\zeta < \infty) = 0$ and therefore that (2.13) holds. Lemma (2.12) shows that for each $x \neq y : \xi(t,x) \neq \xi(t,y)$, $t > 0$, a.s. We now have to show that exceptional set does not depend on x and y.

(2.14) LEMMA: *Let*

$$\eta(t,x,y) = \frac{1}{|\xi(t,x)-\xi(t,y)|}, \quad t > 0 \text{ and } x \neq y.$$

Then η *is a.s. a continuous function of* (t,x,y) *for* $t > 0$ *and* $x \neq y$.

Proof. In view of (2.6) and the fact that η is a.s. continuous on $\{(t,x,y): \xi(t,x) \neq \xi(t,y)\}$, we need only to show that for some $p > 2(2n+1)$, one has

(2.15) $$E[|\eta(t,x,y)-\eta(t',x',y')|^p] \leq C_{p,T}(\delta)(|x-x'|^p+|y-y'|^p+|t-t'|^{p/2})$$

for all $T > 0$, $\delta > 0$, $0 \leq t, t' \leq T$ and $|x-x'| \wedge |y-y'| \geq \delta$.

But

$$|\eta(t,x,y) - \eta(t', x', y')|^p$$

$$= \left| \frac{|\xi(t,x) - \xi(t,y)| - |\xi(t', x') - \xi(t',y')|}{|\xi(t,x) - \xi(t,y)|\,|\xi(t',x') - \xi(t',y')|} \right|^p$$

$$\leq 2^{p-1}|\eta(t,x,y)|^p|\eta(t',x',y')|^p(|\xi(t,x)-\xi(t',x')|^p +$$

$$+ |\xi(t,y) - \xi(t',y')|^p).$$

Thus

$$E[|\eta(t,x,y) - \eta(t',x',y')|^p]$$

$$\leq 2^{p-1} E[|\eta(t,x,y)|^{4p}]^{1/4} E[|\eta(t',x',y')|^{4p}]^{1/4}$$

$$\times (E[|\xi(t,x)-\xi(t',x')|^{2p}]^{1/2}+E[|\xi(t,y)-\xi(t',y')|^{2p}]^{1/2})$$

The first factors are easily estimated by (2.7), so long as $|x - x'| \wedge |y - y'| \geq \delta$.

Moreover, by the argument used to derive (2.4), it is easy to check that

(2.16) $E[|\xi(s,x)-\xi(t,y)|^{2p}] \leq C_p(T)(|s-t|^p + |x-y|^{2p})$, $0 \leq s,t \leq T, x,y \in R$

Thus (2.15) has been proved.

(2.17) Exercise. Prove (2.16) for all $1 \leq p < \infty$.

We now want to show that a.s. the map $x \to \xi(t,x)$ is onto for all $t \geq 0$. The idea is that this is certainly true when $t = 0$ and that the map $\xi(t,.)$ is homotopically connected to $\xi(0,.)$. In order to take advantage of these facts, we must show that they continue to hold on the one-point compactification of R^n.

(2.18) LEMMA: *For each* $T > 0$ *and* $p \in R$ *there is a* $C_p(T) < \infty$ *such that*

$$E[(1 + |\xi(t,x)|^2)^p] \leq C_p(T)(1 + |x|^2)^p, \quad 0 \leq t \leq T.$$

Proof. Let $f(z) = (1 + |z|^2)^p$. Then

$$\partial f/\partial z_i = 2p(1 + |z|^2)^{p-1} z_i$$

and

$$\frac{\partial^2 f}{\partial z_i \partial z_j} = 2p(p-1)(1+|z|^2)^{p-2} z_i z_j + 2p\delta_{ij}(1 + |z|^2)^{p-1} .$$

Hence

$$E[(1+|\xi(t,x)|^2)^p] - (1+|x|^2)^p$$

$$= E[\int_o^t \sum_{i=1}^n b^i(\xi(s,x)\frac{\partial f}{\partial z_i}(\xi(s,x))ds] + \frac{1}{2} E[\int_o^t \sum_{i,j=1}^n \sum_{\ell} \sigma_\ell^i(\xi(s,x))\sigma_\ell^j(\xi(s,x)$$

$$\frac{\partial^2 f}{\partial z_i \partial z_j}(\xi(s,x))ds]$$

$$\leq C \int_o^t E[(1+|\xi(s,x)|^2)^p]ds,$$

since

$$\max_{1\leq i\leq n} |b^i(z)| < \max_{1\leq i\leq n} |b^i(0)| + L|z| < 2 \max_{1\leq i\leq n} |b^i(0)| \vee L(1+|z|^2)^{1/2}$$

and similarly

$$\max_{\substack{1\leq i\leq n \\ 1\leq \ell\leq n}} |\sigma_\ell^i(z)| < 2 \max_{\substack{1\leq i\leq n \\ 1\leq \ell\leq n}} |\sigma_\ell^i(0)| \vee L(1+|z|^2)^{1/2} .$$

Hence the desired result follows by Gronwall's inequality.

(2.19) LEMMA: *Let* $\bar{R}^n = R^n \cup \{\infty\}$ *denote the one-point compactification of* R^n. *Define*

$$\eta(t,x) = \begin{cases} \frac{1}{1+|\xi(t,x)|}, & x \in R^n \\ \cdot & \\ 0, & x = \infty . \end{cases}$$

Then η *is a.s. continuous on* $[0,\infty] \times \bar{R}^n$ *into* R^1 .

Proof. Since η is a.s. continuous on $[0,\infty] \times R^n$, if suffices to show that for each $T > 0$ and $\varepsilon > 0$ there is a.s. an $R > 0$ such that $\eta(t,x) < \varepsilon$ if $|x| \geq R$ and $0 \leq t \leq T$. Choose $p > 2(2n+1)$. Then

$$|\eta(t,y) - \eta(s,x)|^p \leq |\eta(t,y)|^p |\eta(s,x)|^p |\xi(t,y) - \xi(s,x)|^p$$

and so

$$E[|\eta(t,y)-\eta(s,x)|^p] \leq E[|\eta(t,y)|^{4p}]^{1/4} \; E[|\eta(s,x)|^{4p}]^{1/4} E[|\xi(t,y)-\xi(s,x)|^{2p}]^{1/2}$$

Since $(1 + |x|) \geq (1 + |x|^2)^{1/2} \geq \frac{1}{2}(1 + |x|)$, we see from (2.15) that

$$E[|\eta(t,y)|^{4p}]^{1/4} \; E[|\eta(s,x)|^{4p}]^{1/4} \leq K_p(T)(1+|x|)^{-p}(1+|y|)^{-p}$$

for $0 \leq s, t \leq T$.

On the other hand, by (2.15),

$$E[|\xi(t,x) - \xi(s,y)|^{2p}]^{1/2} \leq K_p'(T)(|t-s|^{p/2} + |x-y|^p)$$

for $0 \leq s, t \leq T$. Thus

$$E[|\eta(s,x)-\eta(t,y)|^p] \leq K_p''(T)] \left(\frac{|x-y|}{(1+|x|)(1+|y|)}\right)^p + \frac{|t-s|^{p/2}}{(1+|x|)^p(1+|y|)^p}]$$

We now define

$$\hat{\eta}(t,x) = \eta(t, x/|x|^2) \quad \text{for} \quad x \neq 0,$$

and

$$\hat{\eta}(t, 0) = 0.$$

Since η is a.s. continuous on $[0,\infty] \times R^n$, we will be done once we show that $\hat{\eta}$ admits an a.s. continuous version. Noting that if $|y| \geq |x| > 0$:

$$\frac{\left|\frac{x}{|x|^2} - \frac{y}{|y|^2}\right|}{\left(1+\frac{|x|}{|x|}\right)\left(1+\frac{|y|}{|y|}\right)} = \frac{\frac{1}{|y|^2}\left|\left|\frac{y}{x}\right|^2 x-y\right|}{\frac{1}{|x||y|}(1+|x|)(1+|y|)}$$

$$= \frac{\left|\frac{x}{y}\right|\left|\left|\frac{y}{x}\right|^2 x-x\right| + \left|\frac{x}{y}\right||x-y|}{(1+|x|)(1+|y|)}$$

$$= \frac{\left|\frac{x}{y}\right|^2\left(\left|\frac{y}{x}\right|^2 - 1\right) + \left|\frac{x}{y}\right||x-y|}{(1+|x|)(1+|y|)}$$

$$= \frac{\frac{1}{|y|}(|y| - |x|)\,(|y| + |x|) + \left|\frac{x}{y}\right|\,|x-y|}{(1+|x|)\,(1+|y|)}$$

$$\leq 3|x-y| .$$

We see that

$$E[|\hat{\eta}(s,x) - \hat{\eta}(t,y)|^p] \leq K_p''(T).[|x-y|^p + |t-s|^{p/2}]$$

if $0 \leq s, t \leq T$ and $x,y \neq 0$. Since by (2.18)

$\eta(u,z) \to 0$ in L^p for each $u > 0$ as $|z| \to \infty$, this inequality continues to hold even when x or y is 0. Thus, by (2.6), $\hat{\eta}$ admits an a.s. continuous version. This proves the lemma.

(2.20) <u>THEOREM:</u> *Let* $\sigma(.)$ *and* $b(.)$ *be as in* (2.1) *and let* $\xi(t,x)$ *be as in* (2.11). *Then* a.s. $\xi(t,.)$ *is a homeomorphism of* R^n *onto* R^n *for all* $t \geq 0$.

<u>Proof</u>: We have seen that a.s. $\xi(t.)$ determines a one-to-one continuous map $\bar{\xi}(t,.)$ of $\bar{R}^n$ into itself for all $t \geq 0$. Moreover, $\bar{\xi}(0,.)$ is certainly onto. Thus, by standard homotopy theory, a.s. $\bar{\xi}(t,.)$ is onto for all $t \geq 0$. Also, by the invariance domain theorem $\bar{\xi}(t,.)^{-1}$ must be continuous.

Finally, since a.s. $\bar{\xi}(t,\infty) = \infty$ for all $t \geq 0$, we see that a.s. $\xi(t,.)$ is a homeomorphism of R^n onto R^n for all $t \geq 0$.

§3. Differentiation with respect to x :

We now want to differentiate $\xi(t,x)$ with respect to x.

(3.1) LEMMA: *Let* $\gamma(t,X)$, $t \geq 0$ *and* $X \in R^D$, *be an a.s. continuous* R^M- *valued stochastic process such that* $\gamma(.,X)$ *is* $F.$ - *progressively measurable for each* $X \in R^D$. *Further assume that for* $T > 0$, $R > 0$ *and* $2 \leq p \leq \infty$ *there is a* $C_p(T,R) < \infty$ *such that*

$$E[\sup_{0 \leq t \leq T} |\gamma(t,X)|^p] < C_p(T,R)(1 + |X|)^p, \ |X| \leq R,$$

and

$$E[\sup_{0 \leq t \leq T} |\gamma(t,X)-\gamma(t,Y)|^p] \leq C_p(T,R)|X-Y|^p, |X|V|Y| \leq R.$$

Let $\hat{\sigma} : R^M \times R^N \to R^N \otimes R^d$ *and* $\hat{b} : R^M \times R^N \to R^N$ *be* C^∞- *functions satisfying*

$$\sup_{(\gamma,\eta) \quad R^M \times R^N} \left\| \frac{\partial \hat{\sigma}}{\partial \eta_j} \left(\gamma, \eta\right) \right\|_{H.S.} V \left| \frac{\partial \hat{b}}{\partial \eta_j} \left(\gamma, \eta\right) \right| < \infty ,$$

for $1 \leq j \leq N$, *and assume that* $\hat{\sigma}(\gamma,0)$ *and* $\hat{b}(\gamma,0)$ *and all their derivatives are slowly increasing. Finally, let* $f: R^D \to R^N$ *be a smooth function with bounded first derivatives. Then for each* $X \in R^D$ *the equation*

$$\eta(t,X) = f(X) + \int_0^t \hat{\sigma}(\gamma(s,X),\eta(s,X))d\beta(s) + \int_0^t \hat{b}(\gamma(s,X),\eta(s,X))ds, \ t \geq 0$$

has precisely one solution $\eta(.,X)$. *Moreover, for* T, $R > 0$ *and* $2 \leq p \leq \infty$ *there exist* $C_p'(T,R) < \infty$ *such that*

$$E[\sup_{0 \leq t \leq T} |\eta(t,X)|^p] \leq C_p'(T,R)(1 + |X|)^p, \ |X| \leq R,$$

and

$$E\Big[\sup_{0 \leq t \leq T} |\eta(t,X)-\eta(t,Y)|^p\Big] < C_p'(T,R)\,|X-Y|^p, \quad |X|\vee|Y| \leq R.$$

In particular, η *admits a version which is a.s. continuous in* (t,X).

Proof. The techniques used to prove this lemma are similar to those introduced in section 2. The details are left as an exercise.

(3.2) LEMMA: *Let everything be as in lemma (3.1). Assume, in addition, that a.s.* $\gamma(t,.) \in C^1(R^D)$ *for all* $t \geq 0$ *and that for all* $T, R > 0$ *and* $2 \leq p \leq \infty$

$$\max_{1 \leq \ell \leq D} E\Big[\sup_{0 \leq t \leq T} \Big|\frac{\partial \gamma}{\partial X_\ell}(t,X)\Big|^p\Big] \leq C_p(T,R)(1+|X|)^p, \quad |X| \leq R,$$

and

$$\max_{1 \leq \ell \leq D} E\Big[\sup_{0 \leq t \leq T} \Big|\frac{\partial \gamma}{\partial X_\ell}(t,X)- \frac{\partial \gamma}{\partial X_\ell}(t,X)\Big|^p\Big] \leq C_p(T,R)\,|X-Y|^p, \quad |X|\vee|Y| \leq R .$$

Then a.s. $\eta(t,.) \in C^1(R^n)$ *for all* $t \geq 0$ *and for* $T,R > 0$ *and* $2 \leq p \leq \infty$ *there exists* $C_p'(T,R)$ *such that*

$$\max_{1 \leq \ell \leq D} E\Big[\sup_{0 \leq t \leq T} \Big|\frac{\partial \eta}{\partial X_\ell}(t,X)\Big|^p\Big] \leq C_p'(T,R)(1 + |X|)^p, \quad |X| \leq R,$$

and

$$\max_{1 \leq \ell \leq D} E\Big[\sup_{0 \leq t \leq T} \Big|\frac{\partial \eta}{\partial X_\ell}(t,X) - \frac{\partial \eta}{\partial X_\ell}(t,Y)\Big|^p\Big] \leq C_p'(T,R)\,|X-Y|^p, \quad |X|\vee|Y| \leq R.$$

In particular, for each $1 \leq \ell \leq D$, $\frac{\partial \eta}{\partial X_\ell}$ *admits an a.s. continuous version.*

Proof. Choose and fix $1 \leq \ell \leq D$ and let e_ℓ denote the unit vector in the ℓ-th direction. For $h \in R \setminus \{0\}$, define

$$\Delta_h \eta(t,X) = \eta(t,X + he_\ell) - \eta(t,X),$$

$$\Delta_h \gamma(t,X) = \gamma(t,X + he_\ell) - \gamma(t,X)$$

and

$$\Delta_h f(X) = f(X + he_\ell) - f(X).$$

Then $\Delta_h \eta(.,X)$ is determined by the equation:

$$\Delta_h \eta(t,X) = \Delta_h f(X)$$

$$+ \int_o^t \sum_{i=1}^M \left(\int_o^1 \frac{\partial\hat{\sigma}}{\partial\gamma_i}(\gamma(s,X)+\Theta\Delta_h\gamma(s,X),\eta(s,X)+\Theta\Delta_h\eta(s,X))d\Theta\right)\Delta_h\gamma^i(s,X)d\beta(s)$$

$$+ \int_o^t \sum_{j=1}^N \left(\int_o^1 \frac{\partial\hat{\sigma}}{\partial\eta_j}(\gamma(s,X)+\Theta\Delta_h\gamma(s,X),\eta(s,X)+\Theta\Delta_h\eta(s,X))d\Theta\right)\Delta_h\eta^j(s,X)d\beta(s)$$

$$+ \int_o^t \sum_{i=1}^M \left(\int_o^t \frac{\partial\hat{b}}{\partial\gamma_i}(\gamma(s,X)+\Theta\Delta_h\gamma(s,X),\eta(s,X)+\Theta\Delta_h\eta(s,X))d\Theta\right)\Delta_h\gamma^i(s,X)ds$$

$$+ \int_o^t \sum_{j=1}^N \left(\int_o^t \frac{\partial\hat{b}}{\partial\eta_j}(\gamma(s,X)+\Theta\Delta_h\gamma(s,X),\eta(s,X)+\Theta\Delta_h\eta(s,X))d\Theta\right)\Delta_h\eta^j(s,X)ds$$

Thus if

$$\tilde{\gamma}(t,X,h) = \begin{bmatrix} \tilde{\gamma}_{(0)}(t,X,h) \\ \tilde{\gamma}_{(1)}(t,X,h) \\ \tilde{\gamma}_{(2)}(t,X,h) \\ \tilde{\gamma}_{(3)}(t,X,h) \\ \tilde{\gamma}_{(4)}(t,X,h) \end{bmatrix} = \begin{bmatrix} \gamma(t,X) \\ \Delta_h\gamma(t,X) \\ \frac{1}{h}\Delta_h\gamma(t,X) \\ \eta(t,X) \\ \Delta_h\eta(t,X) \end{bmatrix}$$

where $\frac{1}{h}\Delta_h\gamma(t,X) \equiv \frac{\partial\gamma}{\partial X_\ell}(t,X)$ for $h = 0$, and

$$\tilde{\sigma}(\tilde{\gamma},\tilde{\eta}) = \sum_{i=1}^M \int_o^1 \frac{\partial\sigma}{\partial\gamma_i}(\tilde{\gamma}_{(0)} + \Theta\tilde{\gamma}_{(1)}, \tilde{\gamma}_{(3)} + \Theta\tilde{\gamma}_{(4)})d\Theta\tilde{\gamma}^i_{(2)}$$

$$+ \sum_{j=1}^N \int_o^1 \frac{\partial\hat{\sigma}}{\partial\eta_j}(\tilde{\gamma}_{(0)} + \Theta\tilde{\gamma}_{(1)}, \tilde{\gamma}_{(3)} + \Theta\tilde{\gamma}_{(4)})d\Theta\,\tilde{\eta}^j$$

and $\tilde{b}(\tilde{\gamma},\tilde{\eta})$ is defined analogously, then the process $\tilde{\eta}(t,X,h)$ determined by

$$\widetilde{\eta}(t,X,h) = \frac{\Delta_h f}{h}(X) + \int_0^t \widetilde{\sigma}(\widetilde{\gamma}(s,X,h), \widetilde{\eta}(s,X,h))d\beta(s)$$

$$+ \int_0^t \widetilde{b}(\widetilde{\gamma}(s,X,h), \widetilde{\eta}(s,X,h))ds,$$

where $\Delta_h f/h\ (X) \equiv \frac{\partial f}{\partial X_\ell}(X)$ if $h = 0$, has the property that

$$\widetilde{\eta}(.,X,h) = \Delta_h \eta(.,X)/h \quad \text{a.s.}$$

for each $h \in R \setminus \{0\}$ and $X \in R^D$.

Since, by (3.1), $\widetilde{\eta}$ has an a.s. continuous version, we conclude that $\partial\eta/\partial X_\ell$ exists a.s. and has an a.s. continuous version. Also, by (3.1), $\partial\eta/\partial X_\ell$ satisfies the asserted moment inequalities.

(3.3) <u>THEOREM</u>: *Let* $\sigma: R^n \to R^n \otimes R^d$ *and* $b: R^n \to R^n$ *be* C^∞*-functions such that* $(\partial^{|\alpha|}\sigma/\partial x^\alpha)$ *and* $(\partial^{|\alpha|}b/\partial x^\alpha)$ *are bounded for each* $|\alpha| \geq 1$. *Let* $\xi(t,x)$ *be an a.s. continuous version of the solution to* (2.3). *Then a.s.* $\xi(t,.) \in C^\infty(R^n)$ *for all* $t \geq 0$ *and* $(\partial^{|\alpha|}\xi/\partial x^\alpha)$ *is a.s. continuous in* (t,x) *for all* α. *Moreover, for each* $T, R > 0$, $m \geq 0$ *and* $2 \leq p < \infty$, *there is a* $C_p(T, R, m)$ *such that*

$$\text{(3.4)} \qquad \sup_{|x| \leq R} \sum_{|\alpha| \leq m} E\left[\sup_{0 \leq t \leq T} \left|\frac{\partial^{|\alpha|}\xi}{\partial x^\alpha}(t,x)\right|^p\right] \leq C_p(T, R, m).$$

Finally, for given $m \geq 1$, *let* $\gamma(.,x)$ *denote* $\left\{\frac{\partial^{|\alpha|}\xi}{\partial x^\alpha}(.,x) : |\alpha| \leq m-1\right\}$ *and let* $\eta(.,x)$ *denote* $\left\{\frac{\partial^{|\alpha|}\xi}{\partial x^\alpha}(.,x) : |\alpha| = m\right\}$. *Then* $\eta(.,x)$ *is related to* $\gamma(.,x)$ *by a stochastic differential equation of the sort described in lemma* (3.2).

In particular, if

(3.5) $$J(t,x) = \left(\left(\frac{\partial \xi^i}{\partial x_j}(t,x)\right)\right)_{1 \leq i,\, j \leq n}$$

then $J(t,x)$ *is determined by the equation*

(3.6) $$J(t,x) = I + \sum_{k=1}^{d} \int_0^t S_{(k)}(\xi(s,x))J(s,x)\,d\beta^k(s) + \int_0^t B(\xi(s,x))J(s,x)\,ds,$$

$$t \geq 0,$$

where

$$S_{(k)} = \left(\left(\frac{\partial \sigma_k^i}{\partial x_j}\right)\right)_{1 \leq i,\, j \leq n} \quad \textit{and} \quad B = \left(\left(\frac{\partial b^i}{\partial x_j}\right)\right)_{1 \leq i,\, j \leq n}.$$

Proof. The facts that a.s. $\xi(t,.) \in C^\infty(R^n)$ for all $t \geq 0$ and that $(\partial^{|\alpha|}\xi)/(\partial x^\alpha)$ is a.s. continuous for all α is derived by induction on $|\alpha|$, using (3.2) at each stage. The induction procedure entails showing at each stage that derivatives of order m are related to those of order $\leq m-1$ by an equation of the sort described in lemma (3.2), and so the described equations are a consequence of (3.2).

(3.7) LEMMA: *Suppose that* $\Lambda(.)$, $S_k(.)$ $(1 < k < d)$, *and* $B(.)$ *are* $F.$ *- progressively measurable* $R^N \times R^N$ *- valued functions such that*

$$\max_{1<k<d} ||S_k(.)||_{H.S.} \vee ||B(.)||$$

is uniformly bounded and $A(.)$ *is an* a.s. *continuous solution to*

$$A(t) = A_0 + \sum_{k=1}^{d} \int_0^t S_k(u)A(u)\,d\beta^k(u) + \int_0^t B(u)\,A(u)\,du, \; t \geq 0,$$

where $A_o \in R^N \times R^N$ *is invertible. Then a.s.* $A(t)$ *is invertible for all* $t \geq 0$ *and* $A^{-1}(.)$ *satisfies*

$$A^{-1}(t) = A_o^{-1} - \sum_{k=1}^{d} \int_o^t A^{-1}(u) S_k(u) d\beta^k(u) + \int_o^t (\sum_{k=1}^{d} A^{-1}(u) S_k^2(u) - A^{-1}(u) B(u)) du$$

$$t \geq 0.$$

Proof: The theorem is proved by defining $M(.)$ by the equation which $A^{-1}(.)$ is supposed to satisfy and then using Itô's formula to check that

$$d(M(t)\ A(t)) = d(A(t)\ M(t)) = 0.$$

(3.8) Exercise: In order to see how one can guess the equation satisfied by $A^{-1}(.)$, suppose that $A^{-1}(.)$ exists and denote it by $M(.)$. Assume that

$$dM(t) = \sum_{k=1}^{d} M(t) \tilde{S}_k(t)\ d\beta^k(t) + M(.)\ \tilde{B}(.)dt$$

and use Itô's formula to find out what $\tilde{S}_k(.)$ and $\tilde{B}(.)$ must be.

(3.9) THEOREM: *Let everything be as in theorem* (3.3). *Then a.s.* $J(t,x)$ *is invertible for all* $(t,x) \in [0,\infty] \times R^n$ *and* $J^{-1}(.,x)$ *determined by*

$$J^{-1}(t,x) = I - \sum_{k=1}^{d} \int_o^t J^{-1}(s,x)\ S_k(\xi(s,x)) d\beta^k(s) \tag{3.10}$$

$$+ \int_o^t (\sum_{k=1}^{d} J^{-1}(s,x) S_k^2(\xi(s,x)) J^{-1}(s,x) - J^{-1}(s,x) B(\xi(s,x))) ds.$$

In particular, a.s. $\xi(t,.)$ *is a* C^∞*- diffeomorphism of* R^n *onto* R^n.

§4. An application to Partial Differential Equations:

Let $\xi(t,x)$ be as in section (2). Then it is well-known (cf. Chapter 4 of [S & V]) that $\xi(t,x)$ is strong Markov in the sense that if τ is a finite $F_.$- stopping time and $f \in B(R^n)$ (the bounded measurable functions on R^n) then for all $x \in R^n$:

(4.1) $$E[f(\xi(\tau + t,x))|F_\tau] = u_f(t, \xi(\tau, x)) \quad \text{a.s.}, \; t \geq 0$$

where

(4.2) $$u_f(t,x) = E[f(\xi(t,x))], \quad (t,x) \in [0,\infty] \times R^n.$$

(4.3) LEMMA: *Given* $x \in R^n$ *and* $R > 0$ *there exist* $A_1(x,R) < \infty$ *and* $0 < A_2(x,R) < \infty$ *depending only on* n, $\sup_{y \in B(x,R)} \|\sigma(y)\|_{H.S.}$ *and* $\sup_{y \in B(x,R)} |b(y)|$ *such that*

$$P(\tau_r \leq h) \leq A_1(x,R) \exp(-A_2(x,R)\, r^2/h)$$

for $0 \leq r \leq R$ *and* $h > 0$, *where*

$$\tau_r = \inf\{t \geq 0 : |\xi(t,x) - x| \geq r\}.$$

Proof: Given $\Theta \in R^n$, define

$$X_\Theta(t) = \exp\left[<\Theta, \xi(t,x)-x-\int_0^t b(\xi(s,x)ds> - \frac{1}{2}\int_0^t <\Theta, \sigma\sigma^*(\xi(s,x)\Theta> ds\right]$$

Then by Itô's formula, $(X_\Theta(t \wedge \tau_R), F_t, P)$ is a martingale for all $R > 0$. Set

$$b_R = \sup_{y \in B(x,R)} |b(y)|, \quad a_R = \sup_{y \in B(x,R)} \|\sigma\sigma^*(y)\|_{H.S.}$$

and define $h_R = b_R/2$. For $\Theta \in S^{n-1}$ (the unit sphere in R^{n-1}), $\lambda > 0$, $0 < r \leq R$ and $0 < h \leq h_R$:

$$P(\sup_{0\leq t\leq h} <\Theta,\xi(t,x)-x> \geq r) < P(\sup_{0\leq t\leq h} <\Theta,\xi(t\wedge\tau_R,x)-x-\int_0^{t\wedge\tau_R} b(\xi(s,x))ds) \geq \frac{r}{2})$$

$$< P(\sup_{0\leq t\leq h} X_{\lambda\Theta}(t\wedge\tau_R) \geq \exp(\frac{\lambda r}{2} - \frac{\lambda^2 h}{2} a_R))$$

$$\leq \exp(-\frac{\lambda r}{2} + \frac{\lambda^2 h}{2} a_R)$$

Taking $\lambda = r/(2ha_R)$, we obtain

$$P(\sup_{0\leq t\leq h} <\Theta,\xi(t,x)-x> \geq r) \leq \exp(-r^2/(8ha_R))$$

and so by choosing θ successively to point along n coordinate axes:

$$P(\sup_{0\leq t\leq h} \max_{1\leq i\leq n} |\xi^i(t,x) - x^i| \geq r) \leq 2n\exp(-r^2/8ha_R)$$

Clearly the required estimate follows from this.

(4.4) <u>LEMMA</u>: *Let* $\sigma(.)$ *and* $b(.)$ *be as in theorem* (3.3) *and define* $\xi(t,x)$ *accordingly. Given* $f \in C_\uparrow^\infty(R^n)$ (*the* C^∞-*functions* f *such that the* $D^\alpha f$ *are slowly increasing for all* α), *define* $u_f(t,x)$ *by* (4.1). *Then* $u_f \in C^\infty([0,\infty)\times R^n)$ *and for each* $T, R > 0$ *and* $m \geq 0$ *there is a* $C(T,R,m) < \infty$ *such that*

$$\max_{2\ell+|\alpha|\leq m} \sup_{\substack{0\leq t\leq T\\ |\alpha|\leq R}} \left| \frac{\partial^\ell}{\partial t^\alpha} \frac{\partial^{|\alpha|} u_f}{\partial x^\alpha}(t,x)\right| \leq C(T,R,m) \max_{|\alpha|\leq m} \left\| \frac{\partial^{|\alpha|} f}{\partial x^\alpha}\right\|_u \tag{4.5}$$

Finally, u_f *is the unique* $u \in C_\uparrow([0,\infty) \times R^n) \cap C^{1,2}((0,\infty)\times R^n)$ *such that*

(4.6) $$\partial u/\partial t = Lu \ , \ t \geqslant 0$$

$$\lim_{t \downarrow 0} u(t.,) = f(.)$$

where

$$L = \frac{1}{2} \sum_{i,j=1}^{n} (\sigma\sigma^{*})^{ij}(x) \frac{\partial^2}{\partial x_i \partial x_j} + \sum_{i=1}^{n} b^{i}(x) \frac{\partial}{\partial x_i}$$

Proof: Differentiating $E[f(\xi(t,x))]$ with respect to x, one sees by induction that

(4.7) $$\frac{\partial^{|\alpha|} u_f}{\partial x^{\alpha}}(t,x) = \sum_{\beta \leq \alpha} E\left[\frac{\partial^{|\beta|} f}{\partial x^{\beta}}(\xi(t,x)) P^{(\alpha,\beta)}(\Xi_{(|\alpha|)}(t,x))\right],$$

in $\left\{ \frac{\partial^{|\beta|}\xi}{\partial x^{\beta}}(t,x) : |\beta| \leqslant |\alpha| \right\}$. Using Itô's formula and the fact that $\left\{ \frac{\partial^{|\beta|}\xi}{\partial x^{\beta}}(.,x) : |\beta| \leqslant |\alpha| \right\}$ satisfies a stochastic differential equation with coefficients in $C_{\uparrow}^{\infty}$, one sees that

(4.8) $$\frac{\partial^{\ell}}{\partial t^{\ell}} \frac{\partial^{|\alpha|} u_f}{\partial x^{\alpha}}(t,x) = \sum_{|\beta| \leqslant |\alpha|+2} E\left[\frac{\partial^{|\beta|} f}{\partial x^{\beta}}(t,x)\, \Phi^{(\sigma,\beta,\ell)}(\Xi_{|\alpha|})(t,x)\right]$$

where the $\Phi^{(\alpha,\beta,\ell)} \in C_{\uparrow}^{\infty}(R^{D|\alpha|})$. From (4.8) and moment estimates in (3.3), it is clear that $u_f \in C^{\infty}([0,\infty) \times R^{n})$ and that (4.5) holds.

We next show that u_f satisfies (4.6). Clearly

$$\lim_{t \downarrow 0} u_f(t.,) = f(.).$$

Moreover, by (4.1)

$$u_f(t+h, x) = E[f(\xi(h+t,x))] = E[u_f(t,\xi(h,x))].$$

Let $\tau = \inf \{s \geqslant 0 : |\xi(s,x) - x| \geqslant 1|\}$. Then

$$E[u_f(t,\xi(h,x))] = E[u_f(t,\xi(h \wedge \tau , x))]$$

$$+ E[u_f(t,\xi(h,x)), \ \tau \leq h].$$

Since $u_f(t,.)$ is slcwly increasing, it follows from (4.3) that

$$1/h \ E[u_f(t,\xi(h,x)), \ \tau \leq h] \to 0 \ \text{ as } \ h \downarrow 0.$$

On the other hand, by Itô's formula

$$\frac{1}{h} (E[u_f(t,\xi(h \wedge \tau , x))] - u(t,x))$$

$$= \frac{1}{h} E[\int_0^{h \wedge \tau} L u_f(t,\xi(s,x))ds]$$

$$\to L u_f(t,x) \ \text{ as } \ h \downarrow 0$$

since $P(\tau > 0) = 1$. Thus we have proved that

$$\frac{\partial u_f}{\partial t} = L u_f, \ t \geq 0.$$

Finally, we must show that if $u \in C_{\uparrow}([0,\infty) \times R^n) \cap C^{1,2}((0,\infty) \times R^n)$ satisfies (4.6), then $u = u_f$. To this end, let

$$\tau_R = \inf \{t \geq 0 : \ |\xi(t,x) - x| \geq R\}.$$

Then by Itô's formula for fixed $T > 0$ we know that

$$(u(T - t \wedge \tau_R , \xi(t \wedge \tau_R, x)), F_t, P)$$

is a martingale, $0 \leq t \leq T$. In particular,

$$u(T,x) = E[u(T - T \wedge \tau_R, \xi(T \wedge \tau_R, x)].$$

Since $\tau_R \uparrow \infty$ as $R \downarrow \infty$ and u is slowly increasing,

$$E[u(T - T \wedge \tau_R, \xi(T \wedge \tau_R, x)] \to E[f(x(T))] \ \text{ as } \ R \uparrow \infty .$$

(4.9) <u>Exercise</u>: Under the condition of theorem (3.3), show that for each $m \geq 0$ and $2 \leq p < \infty$ there exist $A_{m,p}, B_{m,p} < \infty$ and $\lambda_{m,p} > 0$ (depending on m,p,n and the bounds on σ and b and their derivatives) such that

$$(4.10)\quad E\left|\sup_{0 \leq t \leq m}\left(\sum_{|\alpha| \leq m}\left|\frac{\partial^{|\alpha|}\xi}{\partial x^{\alpha}}(t,x)\right|^2\right)^{p/2}\right| \leq A_{m,p}\, e^{B_{m,p}T}(1+|x|)^{\lambda_{m,p}} .$$

Calculate from this that if $f \in C_{\uparrow}^{\infty}(R^n)$, then $u_f \in C_{\uparrow}^{\infty}([0,T] \times R^n)$ for all $T > 0$. Check also that for a given $m \geq 1$ and $2 \leq p < \infty$, $\lambda_{m,p}$ does not depend on $\sigma(.)$ and $b(.)$ while $A_{m,p}$ and $B_{m,p}$ depend only on the bounds on

$$\frac{\partial^{|\alpha|}\sigma}{\partial x^{\alpha}}(.) \quad \text{and} \quad \frac{\partial^{|\alpha|}b}{\partial x^{\alpha}}(.) \text{ for } |\alpha| \leq m .$$

CHAPTER II

THE TRAJECTORIES OF SOLUTIONS TO STOCHASTIC DIFFERENTIAL EQUATIONS.

§1. The Martingale Problem :

In Chapter I we have discussed diffusions from Itô's point of view, that is as solutions to stochastic differential equations. For the purpose of certain applications it is useful to adopt a slightly different point of view.

Let $\Omega = C([0,\infty), R^n)$ and think of Ω as a Polish space with the metric of uniform convergence on compact sets. Denote by $\mathcal{M}$ the Borel field over Ω. Given $t > 0$ and $\omega \in \Omega$, let $x(t,\omega)$ be the position in R^n of ω at time t and set $\mathcal{M}_t = \sigma(x(s) : 0 \leq s \leq t)$. If L is a differential operator of the form

$$(1.1) \qquad L = \frac{1}{2} \sum_{i,j=1}^{n} a^{ij}(x) \frac{\partial^2}{\partial x_i \partial x_j} + \sum_{i=1}^{n} b^i(x) \frac{\partial}{\partial x_i} ,$$

where $a(.)$ and $b(.)$ are continuous functions with values in S_n (the set of symmetric, non-negative definite real nxn matrices) and R^n respectively and if $x \in R^n$, we say that P solves the martingale problem for L starting from x (Notation $P \sim L$ at x) if P is a probability measure on $(\Omega, \mathcal{M})$, $P(x(0) = x) = 1$ and for all $f \in C_0^\infty(R^n)$: $(X_f(t), \mathcal{M}_t, P)$ is a martingale, where

$$X_f(t) \equiv f(x(t)) - \int_0^t Lf(x(s))ds.$$

If for each $x \in R^n$ there is precisely one such P, we say that the martingale problem for L is well-posed.

(1.2) THEOREM: *Let* $\sigma(.)$ *and* $b(.)$ *be as in* (I, 3.3) *and let* L *be given as* (1.1) *with* $a(.) = \sigma\sigma^*(.)$. *Then the martingale problem for* L *is well-posed. In fact, for each* $x \in R^n$, *the unique* $P_x \sim L$ *at* x *coincides with the distribution of* $\xi(.,x)$ *is the solution to* (I, 2.3).

Proof: The first step is the simple remark that, by Itô's formula, if $\xi(.,x)$ is given by (I, 2.3), then $P \sim L$ at x, where P denotes the distribution of $\xi(.,x)$. Thus for each $x \in R^n$ the martingale problem for L admits a solution.

The next step is to show that if $P \sim L$ at x and if $f \in C^\infty([0,T) \times R^n)$ for some $T > 0$, then

$$(f(t \wedge T, x(t)) - \int_0^{t \wedge T} (\partial/\partial s + L) f(s,x(s))ds, \mathcal{M}_t, P)$$

is a martingale. The proof of this fact runs as follows:

Given $0 \leq s < t \leq T$ and $N \geq 1$, let $u_k = s + k/N\ (t-s)$. Then for $A \in \mathcal{M}_s$,

$$E^P[f(t, x(t)) - f(s,x(s)), A]$$

$$= \sum_{k=0}^{n-1} E^P[f(u_{k+1}, x(u_{k+1})) - f(u_k, x(u_k)), A]$$

$$= E^P[1/n \sum_{k=0}^{n-1} \frac{\partial f}{\partial u} (u_k, x(u_{k+1})) + \sum_{k=0}^{n-1} \int_{u_k}^{u_{k+1}} Lf(u_k, x(v))dv, A]$$

$$+ 0(1/n)$$

$$\to E^P[\int_s^t (\partial/\partial u + L)\ f(u,x(u))du, A] \quad \text{as} \quad n \to \infty.$$

Having proved the preceding, the identification of $P \sim L$ at x is quite easy. Namely, let $f \in C_0^\infty(R^n)$ and define $u_f(t,x) = E[f(\xi(t,x))]$ as in (I, 4.1). Then $u_f \in C_b([0, \infty) \times R^n) \cap C^\infty([0,\infty) \times R^n)$ and u_f satisfies (I, 4.6). Given $R > 0$, choose $\eta_R \in C_0^\infty(B(0,2R))$ so that $0 \leqslant \eta_R \leqslant 1$ and $\eta_R \equiv 1$ on $\overline{B(0,R)}$. Then the function

$$F_R(t,y) = \eta_R(y)\, u_f(T-t,y) \in C_0^\infty([0,T\,] \times R^n)$$

for any $T > 0$. Hence

$$(F_R(t \wedge T, x(t \wedge T)) - \int_0^t (\partial/\partial s + L)F_R(s,x(s))ds, \mathcal{M}_t, P)$$

is a martingale. Thus by Doob's stopping time theorem,

$$(u_f(T - t \wedge \tau_R \wedge T, x(t \wedge \tau_R \wedge T)), \mathcal{M}_t, P)$$

is a martingale, where

$$\tau_R = \inf \{t \geqslant 0 : |x(t) - x(0)| \geqslant R\}.$$

Since $\tau_R \uparrow \infty$ as $R \uparrow \infty$ and u_f is bounded, it follows that

$$(u_f(T - t \wedge T, x(t \wedge T)), \mathcal{M}_t, P)$$

is a martingale. In particular, if $0 \leqslant s \leqslant T$ then

$$E^P[f(x(T)) \mid \mathcal{M}_s] = E^P[u_f(0, x(T)) | \mathcal{M}_s]$$

$$= u_f(T - s, x(s)) \qquad \text{(a.s., P)}.$$

Working by induction on N, it follows easily that if $0 \leq t_1 < \ldots < t_N$ and $f_1, \ldots, f_N \in C(R^n)$, then

$$E^P[f_1(x(t_1)) \ldots f_N(x(t_N))] = E[f_1(\xi(t_1,x)) \ldots f_N(\xi(t_N,x))]$$

Clearly this identifies P as the distribution of $\xi(\cdot,x)$.

(1.3) <u>Exercise</u>: Using exercise (I, 4.9), show that theorem (1.2) continues to hold if $\sigma(\cdot), b(\cdot) \in C^3(R^n)$ and $(\partial^{|\alpha|} \sigma/\partial x^\alpha)(\cdot)$ and $(\partial^{|\alpha|} b/\partial x^\alpha)(\cdot)$ are bounded for $1 \leq |\alpha| \leq 3$. The point is that under these conditions, one can show that the u_f defined by (I, 4.1) is in $C^{1,2}([0, \infty] \times R^n)$ and still satisfies (I, 4.5). Actually, theorem (1.2) is true even if $\sigma(\cdot)$ and $b(\cdot)$ just satisfy (I, 2.1); however, the proof in this case is quite different (cf. Chapter 5 and 6 of [S & V]).

(1.4) <u>Exercise</u>: Let $\{\sigma_m(\cdot)\}_{m=1}^\infty$ and $\{b_m(\cdot)\}_{m=1}^\infty$ be sequences of co-efficients satisfying the hypotheses of theorem (I, 3.3). Assume that $\sigma_m(\cdot) \to \sigma(\cdot)$ and that $b_m(\cdot) \to b(\cdot)$ uniformly on compact sets where $\sigma(\cdot)$ and $b(\cdot)$ again satisfy the hypotheses of (I, 3.3). For $m \geq 1$ and $x \in R^n$ let $P_x^m \sim L_m$ at x, where L_m is defined for $\sigma_m(\cdot)$ and $b_m(\cdot)$; and for $x \in R^n$ let $P_x \sim L$ at x where L is defined for $\sigma(\cdot)$ and $b(\cdot)$. Show that if $x_m \to x$, then $P_{x_m}^m \Rightarrow P_x$ on Ω, where "$\Rightarrow$" means convergence in the sense of weak convergence of measures. The idea is the following.

In the first place one can easily check that

$$\sup_{m \geq 1} E^{P^m_{x_m}}[|x(t)-x(s)|^4] \leq C(T)|t-s|^2, \quad 0 \leq s \leq t \leq T$$

for each $T > 0$.

Combining this with the fact that $\{x_m\}_1^\infty$ is relatively compact, one can use (I, 2.7) to see that for each $\varepsilon > 0$ there is a compact $K_\varepsilon \subseteq \Omega$ such that $P^m_{x_m}(K_\varepsilon) \geq 1 - \varepsilon$ for all $m \geq 1$. By Prohorov's theorem (cf. Chapter 1 of [S & V]) this means that $\{P^m_{x_m}\}_{m=1}^\infty$ is relatively compact in the weak topology. Finally, one can easily check that if $\{P^{m'}_{x_{m'}}\}$ is any weakly convergent subsequence of $\{P^m_{x_m}\}$ and $P^{m'}_{x_{m'}} \Rightarrow P$, then $P \sim L$ at x; and so $P^m_{x_m} \Rightarrow P_x$.

§2. APPROXIMATING DIFFUSIONS BY RANDOM EVOLUTIONS

We start with an example:

Let $\tau_1, \tau_2, \ldots, \tau_n \ldots$ be independent unit exponential random variables on some probability space (i.e. $P(\tau_1 > s_1, \ldots, \tau_n > s_n) = \exp(-\sum_{j=1}^n s_j)$ for all $(s_1, \ldots, s_n)$ $[0,\infty)^n$). Define $T_0 \equiv 0$, and $T_n = \sum_{j=1}^n \tau_j$, $n \geq 1$; and consider $N(t) = \max\{n \geq 0: T_n \leq t\}$. Since $P((\exists\, n > 1): \tau_n = 0) = 0$, it is clear that a.s. the path $t \to N(t)$ is right continuous, non-decreasing, piecewise constant and jumps by 1 when it jumps.

(2.1) <u>LEMMA</u>: *For* $0 \leq s < t$ *and* $m \geq 0$:

(2.2) $$P(N(t) - N(s) = m | \sigma(N(u),\, u \leq s)) = \frac{(t-s)^m}{m!} e^{-(t-s)} \quad \text{a.s.}$$

That is, $N(t) - N(s)$ *is independent of* $\sigma(N(u): 0 \leq u \leq s)$ *and is a Poisson random variable with intensity* $(t-s)$.

<u>Proof</u>: Note that if $t \geq 0$ and $h > 0$,

$$P(N(t+h) - N(t) \geq 1 / \sigma(N(u) : u \leq t))$$

$$= P(N(t+h) - N(t) \geq 1 / \sigma(T_1, \ldots, T_{N(t)}, N(t)))$$

$$= P(\tau_{N(t)+1} \leq t+h-T_{N(t)} / \sigma(T_1, \ldots, T_{N(t)}, N(t)))$$

$$= P(\tau_{N(t)+1} \leq t+h-T_{N(t)} / \tau_{N(t)+1} > t - T_{N(t)})$$

$$= 1 - P(\tau_{N(t)+1} > t+h - T_{N(t)} / \tau_{N(t)+1} > t - T_{N(t)})$$

$$= 1 - e^{-h}$$

$$= h + \phi(h)$$

Similarly

$$P(N(t+h) - N(t) \geq 2/\sigma(N(u) : u \leq t))$$

$$= P(\tau_{N(t)+1} + \tau_{N(t)+2} \leq t+h - T_{N(t)} / \tau_{N+1} > t - T_{N(t)})$$

$$= \int_0^h e^{-u}(1 - e^{-(h-u)})du = \psi(h).$$

Because of the structure of the paths $N(.)$, if $0 \leq s < t$ and $u_{n,k} \equiv s + \frac{k}{n}(t-s)$ $(n \geq 1$ and $0 \leq k \leq n)$, then we now see that

$$P(N(t) - N(s) = m | \sigma(N(u) : u \leq s))$$

$$= \lim_{n \to \infty} \sum_{\substack{A \subset \{1, \ldots, n\} \\ |A| = m}} P(N(u_{n,k}) - N(u_{n,k-1}) = \chi_A(k),\ 1 \leq k \leq n | \sigma(N(u) : u \leq s))$$

$$= \lim_{n \to \infty} \binom{n}{m} (h_n + \phi(h_n) - \psi(h_n))^m (1 - h_n - \phi(h_n))^{n-m}$$

$$= \frac{(t-s)^m}{m!} e^{-(t-s)}$$

where we have used $h_n \equiv \frac{t-s}{n}$

Next define $\theta(t) = (-1)^{N(t)}$, $t \geq 0$. Clearly $\theta(.)$ is right continuous, piece-wise constant and $\theta(t) \neq \theta(t-)$ implies $\theta(t) = -\theta(t-)$.

(2.3) <u>LEMMA:</u> *If* $f: R^1 \times \{-1, 1\} \to R^1$ *is a function such that* $f(.,\theta) \in C_b^1(R')$, θ $\{-1,1\}$, *then*

$$\left(f\left(\int_0^t \theta(u)du, \theta(t)\right) - \int_0^t \left[\theta(s) f'\left(\int_0^t \theta(u)du, \theta(s)\right) + Kf\left(\int_0^s \theta(u)du, \theta(s)\right)\right]ds,\ F_t,\ P\right)$$

is a martingale, where $F_t = \sigma(\theta(u): 0 \leq u \leq t)$, $f'(x,\theta) = \frac{d}{dx} f(x,\theta)$ *and* $Kf(x,\theta) = f(x, -\theta) - f(x,\theta)$.

<u>Proof</u>: If we can prove the result when f does not depend on x, then the general case follows immediately by the argument given in the second step of the proof of theorem (II, 4.2). But from (2.2) it is easy to see that

$$\frac{d}{dt} E[f(\theta(t))|F_s = E[Kf(\theta(t))|F_s\ ,\ 0 \leq s \leq t,$$

and so the result holds for f not depending on x.

Consider the process

(2.4) $$x_\varepsilon(t) = \varepsilon \int_0^t \theta(s)\, ds,$$

where $\varepsilon > 0$, and define

(2.5) $$X_\varepsilon(t) = x_\varepsilon(t) + \frac{\varepsilon}{2}\theta(t).$$

(2.6) <u>LEMMA</u>: *If* $\phi \in C_0^\infty(R^1)$, *then there is an* $F.$*- progressively measurable process* $L_{\phi,\varepsilon}(.)$ *such that* $|L_{\phi,\varepsilon}(.)| < C\|\phi\|_{C_b^3(R^1)}$ *and*

$$\left(\phi(X_\varepsilon(t)) - \frac{\varepsilon^2}{2}\int_0^t \phi''(X_\varepsilon(s))\, ds - \varepsilon^3 \int_0^t L_{\phi,\varepsilon}(s) ds,\ F_t,\ P\right)$$

is a martingale.

Proof: By (2.3) with $f(x,\theta) = \phi(\varepsilon x + \frac{\varepsilon}{2}\theta)$:

$$(\phi(X_\varepsilon(t)) - \varepsilon\int_0^t \theta(s)\phi'(X_\varepsilon(s))ds - \int_0^t (\phi(X_\varepsilon(s)-\varepsilon\theta(s))-\phi(X_\varepsilon(s)))ds,\ F_t,\ P)$$

is a martingale. By Taylor's theorem

$$\phi(X_\varepsilon(s) - \varepsilon\theta(s)) - \phi(X_\varepsilon(s))$$
$$= \varepsilon\theta(s)\phi'(X_\varepsilon(s)) + \frac{\varepsilon^2}{2}\phi''(X_\varepsilon(s)) - \frac{\varepsilon^2}{6}\phi'''(X_\varepsilon(s)-\delta(s)\theta(s))$$

where $0 < \delta(s) < \varepsilon$. Thus we can take

$$L_{\phi,\varepsilon} = -\frac{1}{6}\phi'''(X_\varepsilon(s) - \delta(s)\theta(s)).$$

We now see that for $\phi \in C_0(R^1)$:

$$(\phi(X_\varepsilon(\frac{t}{\varepsilon^2})) - \frac{1}{2}\int_0^t \phi''(X_\varepsilon(\frac{s}{\varepsilon^2}))ds - \varepsilon\int_0^t L_{\phi,\varepsilon}(\frac{s}{\varepsilon^2})ds,\ F_{t/\varepsilon^2},\ P)$$

is a martingale. Thus if we let P^ε be the distribution on $\Omega(n = 1)$ of $X_\varepsilon(\cdot/\varepsilon^2)$, then, since $|x_\varepsilon(\cdot/\varepsilon^2) - X_\varepsilon(\cdot/\varepsilon^2)| \leqslant \varepsilon/2$, it is reasonable to suppose that $P^\varepsilon \Rightarrow \omega$ as $\varepsilon \downarrow 0$, where $\omega \sim \frac{1}{2}\frac{d^2}{dx^2}$ at 0 (i.e. ω is 1-dimensional Wiener measure), in the weak topology. If we know that $\{P^\varepsilon : \varepsilon > 0\}$ were compact, then this convergence would follow immediately from the observation that if $\varepsilon_n \downarrow 0$ and $P^{\varepsilon_n} \Rightarrow P$, then

$$(\phi(x(t)) - \frac{1}{2}\int_0^t \phi''(x(s))\,ds,\ M_t,\ P)$$

is a martingale for all $\phi \in C_0^\infty(R^1)$.

In order to handle the compactness question, we state the following theorem.

(2.7) <u>THEOREM:</u> *Let* $\{P^{\varepsilon}: \varepsilon > 0\}$ *be a family of probability measures on* $(\Omega, \ )$ *and let* $A(\phi)$, $\phi \in C_0^{\infty}(R^n)$, *be a non-negative number such that* $A(\phi) < C\|\phi\|_{C^k(R^n)}$ *for some* $C < \infty$ *and* $k \geq 0$. *Assume that for each* $\varepsilon > 0$ *there is a* $M.$*-progressively measurable function* $X_{\varepsilon}: [0,\infty) \times \Omega \to R^n$ *such that*

(i) $\lim\limits_{R \uparrow \infty} \sup\limits_{\varepsilon > 0} P^{\varepsilon}(|X_{\varepsilon}(0)| \geq R) = 0$

(ii) $X_{\varepsilon}(.)$ *is right continuous and has left limits* $(P^{\varepsilon}$- a.s.),

(iii) $\overline{\lim\limits_{\varepsilon}} P^{\varepsilon}(\operatorname{Sup}\limits_{0 \leq t \leq T} |x(t) - X_{\varepsilon}(t)| \geq \delta) = 0$ *for each* $T > 0$ *and* $\delta > 0$,

(iv) $(\phi(X_{\varepsilon}(t)) + A(\phi)t, M_t, P^{\varepsilon})$ *is a submartingale for each* $\varepsilon > 0$.

If $\{\varepsilon_n\}_1^{\infty}$ *is any sequence of positive numbers tending to zero, then* $\{P^{\varepsilon_n}: n \geq 1\}$ *has a weakly convergent sub-sequence.*

<u>Comment on the proof</u>: We will not give the proof here because it is somewhat involved. However, in the case when $X_{\varepsilon}(.) = x(.)$ for all $\varepsilon > 0$, the proof is given in Chapter 1 of [S & V] (cf. Theorem 1.4.6) and the proof of the general case can be easily accomplished by modifying the lemma 1.4.1 given there. The modification is the following.

<u>LEMMA:</u> *Let* t_1 *and* t_2 *be any pair of points in* $[0,T]$, *such that* $|t_2 - t_1| < \delta_{\omega}(\rho)$. *Then*

$$|X(t_2,\omega) - X(t_1,\omega)| \leqslant \rho + \sup_{0 \leqslant t \leqslant T} |X_\varepsilon(t,\omega) - x_\varepsilon(t,\omega)|$$

and so

$$\sup\{|X(t_2,\omega)-X(t_1,\omega)| : 0 \leqslant t_1 \leqslant t_2 \leqslant T \quad \textit{and} \quad |t_2-t_1| < \delta(\rho)\}$$

$$< \rho + 2 \sup_{0 \leqslant t \leqslant T} |X_\varepsilon(t,\omega) - x_\varepsilon(t,\omega)|$$

The notation is the same as that in lemma 1.4.1 of [S & V]. We now see that the measures P^ε discussed in the paragraph preceding (2.7) converge weakly, as $\varepsilon \downarrow 0$, to ω. Notice that this convergence result provides some insight into the structure of Brownian paths. Indeed, the paths of $x_\varepsilon(./\varepsilon^2)$ are rather simple; they all have speed $1/\varepsilon$ and the times at which they change directions are distributed like sums of independent exponential random variables having mean ε^2. In the limit, this constant speed property is reflected by the a.s. constancy of the square variation of Brownian paths over a given time interval.

We now want to generalize the preceding in order to get analogous approximations of more general diffusions. To this end, let $G = SO(d)$ and let λ denote normalized Haar measure on G. By expanding the original probability space on which the τ_j's were defined, we may assume that there exist G-valued random variables $g_1, g_2, \ldots, g_n, \ldots$ which are independent of τ_j's and independent of one another, and each have distribution λ. Let $\theta_0 \in S^{d-1}$ be fixed and define $\theta(t) = g_{N(t)}\, g_{N(t)-1} \cdots g_1\, \theta_0 (\equiv \theta_0$ if $N(t) = 0)$. Set $F_t = \sigma(\theta(u): 0 \leqslant u \leqslant t)$. By the same argument as that used to prove (2.3), one can prove that if

$$A(t) = \int_0^t \gamma(s)ds$$

where $\gamma(.)$ is a bounded, right continuous, R^N-valued, $F.$-progressively measurable function and if $f \in C_b^{1,0}(R^N \times S^{d-1})$, then

(2.8) $$(f(A(t),\theta(t)) - \int_0^t [< \gamma(s), \mathrm{grad}_x\, f(\Lambda(s),\theta(s))> + Kf(A(s),\theta(s))]ds, F_t, P)$$

is a martingale; where

$$Kf(x,\theta) = \int_{S^{n-1}} (f(x,\eta) - f(x,\theta))d\eta$$

and $d\eta$ denotes normalized uniform measure on S^{n-1}.

(2.9) <u>LEMMA</u>: *For each* $\varepsilon > 0$, *let* $\gamma_\varepsilon(.)$ *be a right continuous* R^n- *valued,* $F.$- *progressively measurable function such that*

$$|\gamma_\varepsilon(.)| \leq \rho(\varepsilon)$$

where $\rho(\varepsilon) \downarrow 0$ *as* $\varepsilon \downarrow 0$. *Set*

$$A_\varepsilon(t) = \int_0^t \gamma_\varepsilon(s)ds.$$

Then for any $h \in C^{1,0}(R^n \times S^{d-1})$, $T > 0$ *and* $\delta > 0$:

$$\lim_{\varepsilon \downarrow 0} P(\int \sup_{0 \leq t \leq T} |\varepsilon \int_0^{t/\varepsilon} (h(A_\varepsilon(s),\theta(s)) - \bar{h}(A_\varepsilon(s)))ds \geq \delta|) = 0$$

where $\bar{h}(x) \equiv \int_{S^{d-1}} h(x,\eta)d\eta$.

<u>Proof</u>: Clearly we may assume that $h(0,\theta_o) = 0$. Set

$$\Delta_\varepsilon(t) = \varepsilon \int_0^t h(A_\varepsilon(s),\theta(s)) - \bar{h}(A_\varepsilon(s)))ds$$

and $$\tilde{\Delta}_\varepsilon(t) = \Delta_\varepsilon(t) + \varepsilon h(A_\varepsilon(t), \theta(t)).$$

Clearly it suffices to prove that

$$\sup_{0\le t\le T} |\tilde{\Delta}_\varepsilon(t/\varepsilon)| \to 0 \text{ in probability, as } \varepsilon \downarrow 0.$$

Thus it is enough to check that if $\phi \in C_b^2(R^1)$ and $\phi(0) = 0$, then

$$E[\sup_{0\le t\le T} \phi(\tilde{\Delta}_\varepsilon(t/\varepsilon)] \to 0 \text{ as } \varepsilon \downarrow 0.$$

To this end, note that by (2.8)

$$\begin{aligned} M_\varepsilon(t) \equiv \phi(\tilde{\Delta}_\varepsilon(t)) &- \varepsilon \int_0^t (h-\bar{h})\ (A_\varepsilon(s), \theta(s))\ \phi'(\tilde{\Delta}_\varepsilon(s))ds \\ &- \varepsilon \int_0^t < \gamma_\varepsilon(s), \operatorname{grad}_x h(A_\varepsilon(s),\theta(s)) > \phi'(\tilde{\Delta}_\varepsilon(s))ds \\ &- \int_0^t (\int_{S^{n-1}} [\phi(\Delta_\varepsilon(s)+h(A_\varepsilon(s),\eta))-\phi(\Delta_\varepsilon(s)+ \varepsilon h(A_\varepsilon(s),\theta(s))]d\eta)ds \end{aligned}$$

is an $F.$- martingale with respect to P.

By Taylor's theorem

$$\begin{aligned} &\int_{S^{d-1}} [\phi(x+\varepsilon h(A_\varepsilon(s),\eta)) - \phi(x+\varepsilon h(A_\varepsilon(s), \theta(s)))]d\eta \\ &= \varepsilon\phi'(x) \int_{S^{d-1}} [h(A_\varepsilon(s),\eta)- h(A_\varepsilon(s),\theta(s))]d\eta \\ &\quad + \varepsilon^2/2 \int_{S^{d-1}} R_\varepsilon(x, h(A_\varepsilon(s),\eta), h(A_\varepsilon(s), \theta(s))d\eta \end{aligned}$$

where $|R_\varepsilon(.)| < C\|\phi\|_{C_b^2(R^1)}$. Thus

$$\sup_{0\le t\le T} |\phi(\tilde{\Delta}_\varepsilon(t/\varepsilon))- M_{\varepsilon,\phi}(t/\varepsilon)| \le C(\phi)(\varepsilon+\delta(\varepsilon))T$$

and so we need only to show that $E[\sup_{0\le t\le T} M^2_{\varepsilon,\phi}(t/\varepsilon)] \to 0$ as $\varepsilon \downarrow 0$. But $E[\sup_{0\le t\le T} M^2_{\varepsilon,\phi}(t/\varepsilon)] \le 4E[M^2_{\varepsilon,\phi}(T/\varepsilon)]$, and

$$M^2_{\varepsilon,\phi}(T/\varepsilon) \leq 2\phi^2(\tilde{\Delta}_\varepsilon(T/\varepsilon)) + 2(C(\phi))^2 (\varepsilon+\delta(\varepsilon))^2 T^2 .$$

Thus it suffices to show that $E[\phi^2(\tilde{\Delta}_\varepsilon(T/\varepsilon))] \to 0$ as $\varepsilon \downarrow 0$.

Finally,

$$|\phi^2(\tilde{\Delta}_\varepsilon(T/\varepsilon)) - M_{\varepsilon,\phi^2}(T/\varepsilon)| \leq C(\phi^2)\ (\varepsilon + \delta(\varepsilon))T,$$

and so

$$E[\phi^2(\tilde{\Delta}_\varepsilon(T/\varepsilon)) \leq C(\phi^2)\ (\varepsilon + \delta(\varepsilon))T,$$

since

$$E[M_{\varepsilon,\phi^2}(T/\varepsilon)] = E[M_{\varepsilon,\phi^2}(0)] = 0.$$

(2.10) <u>THEOREM:</u> *Let* $F : R^n \times S^{d-1} \to R^n$ *be a smooth bounded function having bounded derivatives and satisfying*

$$\int_{S^{d-1}} F(x,\eta)d\eta = 0 \quad \textit{for all } x \in R^n.$$

Let $b : R^n \to R^n$ *be a smooth bounded function having bounded derivatives. Define*

$$a^{ij}(x) = \int_{S^{d-1}} \int_{S^{d-1}} (F^i(x,\eta)-F^i(x,\xi))(F^j(x,\eta)-F^j(x,\xi))d\xi\ d\eta$$

and

$$c^i(x) = \int_{S^{d-1}} \sum_{j=1}^n F^j(x,\eta) \frac{\partial F^i}{\partial x_j}(x,\eta)d\eta\ ;$$

and define $x_\varepsilon(\cdot,x)$ *by*

$$x_\varepsilon(t,x) = x+\varepsilon \int_0^t F(x_\varepsilon(s,x),\theta(s))ds + \varepsilon^2 \int_0^t b(x_\varepsilon(s,x),\theta(s)\ ds,\ t \geq 0,$$

for $\varepsilon > 0$ *and* $x \in R^n$.

Let $P_x^{(\varepsilon)}$ *on* (Ω, M) *be the distribution of* $x_\varepsilon(./\varepsilon^2)$ *and set*

$$L = \frac{1}{2} \sum_{i,j=1}^{n} a^{ij}(x) \frac{\partial^2}{\partial x_i \partial x_j} + \sum_{i=1}^{n} (b^i + c^i)(x) \frac{\partial}{\partial x_i}$$

Then $\{P_x^{(\varepsilon)} : 0 < \varepsilon \leq 1\}$ *is relatively compact in the weak topology and if* $P_x^{(\varepsilon_n)} \Rightarrow P$ *where* $\varepsilon_n \downarrow 0$ *then* $P \sim L$ *at* x.

Proof: Without loss of generality, we will assume that $x = 0$ and for convenience we will use $x_\varepsilon(.)$ in place of $x_\varepsilon(.,0)$.

Define

$$X_\varepsilon(t) = x_\varepsilon(t) + \varepsilon F(x_\varepsilon(t), \theta(t)).$$

Given $\phi \in C_0^\infty(R^n)$, set

$$\tilde{\phi}(x,\theta) = \phi(x + \varepsilon F(x,\theta)).$$

Then by (2.8):

$$\phi(X_\varepsilon(t)) - \int_0^t < \varepsilon F(x_\varepsilon(s), \theta(s)) + \varepsilon^2 b(x_\varepsilon(s)), \operatorname{grad}_x \tilde{\phi}(x_\varepsilon(s), \theta(s)) > ds$$
$$- \int_0^t \left(\int_{S^{d-1}} [\tilde{\phi}_\varepsilon(x_\varepsilon(s), \eta) - \tilde{\phi}_\varepsilon(x_\varepsilon(s), \theta(s))] d\eta\right) ds$$

is an $F.$-martingale with respect to P. Note that

$$\frac{\partial \tilde{\phi}}{\partial x_i}(x,\theta) = \frac{\partial \phi}{\partial x_i}(x + \varepsilon F(x,\theta)) + \varepsilon \sum_{j=1}^{n} \frac{\partial \phi}{\partial x_j}(x + \varepsilon F(x,\theta)) \frac{\partial F^j}{\partial x_i}(x,\theta).$$

Next, observe that by Taylor's theorem

$$\int_{S^{d-1}} [\tilde{\phi}_\varepsilon(x,\eta) - \tilde{\phi}_\varepsilon(x,\theta)] d\eta$$
$$= \varepsilon \int_{S^{d-1}} < F(x,\eta) - F(x,\theta), \operatorname{grad}_x \phi(x + \varepsilon F(x,\theta)) > d\eta$$
$$+ \frac{\varepsilon^2}{2} \sum_{i,j=1}^{n} \int_{S^{d-1}} (F^i(x,\eta) - F^i(x,\theta))(F^j(x,\eta) - F^j(x,\theta)) d\eta \frac{\partial^2 \phi}{\partial x_i \partial x_j}(x + \varepsilon F(x,\theta))$$
$$+ \varepsilon^3 R_{\varepsilon,\phi}(x,\theta)$$

where

$$|R_{\varepsilon,\phi}(x,\theta)| \le C\|\phi\|_{C_b^3(R^n)}$$

Thus if

$$a^{ij}(x,\theta) = \int_{S^{d-1}} (F^i(x,\eta) - F^i(x,\theta))(F^j(x,\eta) - F^j(x,\theta))d\eta$$

and

$$\beta^i(x,\theta) = b^i(x) + \sum_{j=1}^{n} (F^j \frac{\partial F^i}{\partial x_j})(x,\theta) ,$$

then, since

$$\int F(x,\eta)\, d\eta = 0,$$

$$\phi(X_\varepsilon(t)) - \varepsilon^2 \int_0^t < \beta(x_\varepsilon(s), \theta(s)), \operatorname{grad}_x \phi(X_\varepsilon(s)) > ds$$

$$- \frac{\varepsilon^2}{2} \int_0^t \sum_{i,j=1}^{n} a^{ij}(x_\varepsilon(s), \theta(s)) \frac{\partial^2\phi}{\partial x_i \partial x_j} (X_\varepsilon(s))ds$$

$$- \varepsilon^3 \int_0^t L_{\varepsilon,\phi}(x_\varepsilon(s), \theta(s))ds$$

is an F.-martingale with respect to P, where

$$|L_{\varepsilon,\phi}(x,\theta)| \le C'\|\phi\|_{C_b^3(R^n)}$$

By (2.7) this proves that $\{P^\varepsilon : \varepsilon > 0\}$ is relatively compact. Also, it shows that there is an $\tilde{L}_{\varepsilon,\phi}(x,\theta)$ satisfying

$$|\tilde{L}_{\varepsilon,\phi}(x,\theta)| \le C''\|\phi\|_{C_b^3(R^n)}$$

such that

$$\phi(x_\varepsilon(t))-\varepsilon^2 \int_0^t < \beta(x_\varepsilon(s), \theta(s)), \text{grad } \phi(X_\varepsilon(s)) > ds$$

$$- \frac{\varepsilon^2}{2} \int_0^t \sum_{i,j=1}^n a^{ij}(x_\varepsilon(s), \theta(s)) \frac{\partial^2\phi}{\partial x_i \partial x_j}(x_\varepsilon(s))ds$$

$$- \varepsilon^3 \int_0^t \tilde{L}_{\varepsilon,\phi}(x_\varepsilon(s),\theta(s))ds + (\phi(X_\varepsilon(t)) -\phi(x_\varepsilon(t)))$$

is an $F.$-martingale with respect to P. Since

$$|\phi(X_\varepsilon(t)) - \phi(x_\varepsilon(t))| \le C''\varepsilon$$

and, by (2.8), we know that

$$\sup_{0\le t\le T} |\varepsilon^2 \int_0^{t/\varepsilon^2} [L\phi(x_\varepsilon(s)) - < \beta(x_\varepsilon(s), \theta(s)), \text{grad } \phi(x_\varepsilon(s)) >$$

$$- \frac{1}{2} \sum_{i,j=1}^n a^{ij}(x_\varepsilon(s), \theta(s)) \frac{\partial^2\phi}{\partial x_i \partial x_j} (x_\varepsilon(s))]ds|$$

$\to 0$ in probability, it is now clear that $\varepsilon_n \downarrow 0$ and $P^{\varepsilon_n} \Rightarrow P$, then $P \sim L$ at 0.

(2.11) <u>COROLLARY</u>: *Let* $\sigma: R^n \to R^n \otimes R^d$ *and* $b: R^n \to R^n$ *be bounded smooth functions having bounded derivatives of all orders and define*

$$(2.12) \qquad L = \frac{1}{2} \sum_{k=1}^d (\sum_{i=1}^n \sigma_k^i(x) \frac{\partial}{\partial x_i})^2 + \sum_{i=1}^n b^i(x)\frac{\partial}{\partial x_i} .$$

For $\varepsilon > 0$ *and* $x \in R^n$, *let* $x_\varepsilon(.,x)$ *be the process determined by*

$$(2.13) \qquad x_\varepsilon(t,x) = x+ \varepsilon \int_0^t \sigma(x_\varepsilon(s,x))\theta(s)ds+\varepsilon^2 \int_0^t b(x_\varepsilon(s,x))ds, \ t \geq 0,$$

and denote by P_x^ε *the distribution on* $(\Omega, \ )$ *of* $x_\varepsilon(./\varepsilon^2, x)$. *Then as* $\varepsilon \downarrow 0$, P_x^ε *converges weakly to the unique* $P_x \sim L$ *at* x. *In particular, if* $\mathcal{S}(x; \sigma,b)$ *denotes the set of paths* $\phi_\psi(.)$ *of the form*

$$(2.14) \qquad \phi_\psi(t) = x + \int_0^t \sigma(\phi_\psi(s))\psi(s)\, ds + \int_0^t b(\phi_\psi(s))ds ,$$

where $\psi : [0,\infty] \to R^d$ *is a locally bounded, right continuous function possessing left limits, then*

$$\text{supp}(P_x) \subseteq \overline{\mathcal{S}(x;\sigma,b)}$$

<u>Proof</u>: The second part follows immediately from the first, since (2.13) shows explicitly that

$$\text{Supp}(P_x^\varepsilon) \subseteq \overline{\mathcal{S}(x;\ \sigma,\ b)}$$

for each $\varepsilon > 0$.

To prove the convergence result, first observe that, by (1.2), the martingale problem for L is well-posed. Next, define

$$F^i(x,\theta) = (d/2)^{1/2} \sum_{k=1}^{d} \sigma_k^i(x)\theta_k \ , \ 1 \leq i \leq n.$$

Then $F(x,\theta)$ satisfies the hypotheses of theorem (2.10). Moreover,

$$\int_{S^{d-1}} (F^i(x,\theta) - F^i(x,\eta))(F^j(x,\theta) - F^j(x,\eta))d\eta$$

$$= \frac{d}{2} \sum_{k,\ell} \sigma_k^i(x)\ \sigma_\ell^j(x) \int_{S^{d-1}} (\theta_k - \eta_k)\ (\theta_\ell - \eta_\ell)d\eta$$

$$= \frac{d}{2} \sum_{k,\ell} \sigma_k^i(x)\ \sigma_\ell^j(x)\ (\theta_k\ \theta_\ell + \frac{1}{d}\ \delta_{k,\ell}).$$

Thus in the notation of (2.10) :

$$a^{ij}(x) = \sum_\ell \sigma_\ell^i(x)\ \sigma_\ell^j(x).$$

Next:

$$\int_{S^{d-1}} \sum_{j=1}^{n} F^i(x,\eta) \frac{\partial F^i}{\partial x_j}(x,\eta)\,d\eta$$

$$= \frac{d}{2} \sum_{j=1}^{n} \sum_{k,\,\ell=1}^{d} \sigma_k^j(x) \frac{\partial \sigma_\ell^i}{\partial x_j}(x) \int_{S^{d-1}} \eta_k \eta_\ell \, d\eta$$

$$= \frac{1}{2} \sum_{j=1}^{n} \sum_{k=1}^{d} \sigma_k^j(x) \frac{\partial \sigma_k^i}{\partial x_j}(x)$$

Thus in the notation of (2.10)

$$c^i(x) = \frac{1}{2} \sum_{k=1}^{d} \sum_{j=1}^{n} \sigma_k^j(x) \frac{\partial \sigma_k^i}{\partial x_j}(x) .$$

But this means that

$$\frac{1}{2} \sum_{i,j=1}^{n} a^{ij}(x) \frac{\partial^2}{\partial x_i \partial x_j} + \sum_{i=1}^{n} c^i(x) \frac{\partial}{\partial x_i} = \frac{1}{2} \sum_{k=1}^{d} \Big(\sum_{i=1}^{n} \sigma_k^i(x) \frac{\partial}{\partial x_i} \Big)^2 ,$$

and so the L is given in (2.12) is the one associated with $F(x,\theta)$ and $b(x,\theta)$ via the prescription given in (2.10).

(2.15) <u>Exercise</u>: Let $\sigma(.)$ and $b(.)$ be as in the preceding and suppose that $\bar{\sigma}(.)$ and $\bar{b}(.)$ are a second pair of such functions. Assume that Range $(\bar{\sigma}(x)) \subseteq$ Range $(\sigma(x))$ for all $x \in R^n$ and that $\bar{b}(x) - b(x) \in$ Range $(\bar{\sigma}(x))$ for all $x \in R^n$. Show that

$$\overline{\mathcal{S}(x;\, \bar{\sigma},\, \bar{b})} \subseteq \overline{\mathcal{S}(x;\, \sigma,\, b)}$$

for all $x \in R^n$.

(2.16) <u>Remark</u>: An alternative description of $\overline{\mathcal{S}(x;\, \sigma,\, b)}$ can be obtained as follows:

Let

$$X_k = \sum_{i=1}^{n} \sigma_k^i(x) \frac{\partial}{\partial x_i}, \quad 1 < k < d,$$

and denote by Lie $(X_1, \ldots, X_d)$ the Lie algebra of vector fields generated by $\{X_1, \ldots, X_d\}$. Then $\overline{\mathcal{L}(x; \sigma, b)}$ is the closure of the integral curves, starting at x, of vector fields $Z + Y$, where $Z \in$ Lie $(X_1, \ldots, X_d)$ and $Y = \sum_{i=1}^{n} b^i(x) \frac{\partial}{\partial x_i}$. The derivation of this identification rests on the elementary facts about how integral curves change under the Lie bracket operation and linear combinations.

§ 3. CHARACTERIZATION OF SUPP(P_x), THE NON-DEGENERATE CASE:

In Corollary (2.11), we showed that if L is given by (2.12) and $P_x \sim L$ at x, then $\text{Supp}(P_x) \subseteq \overline{\mathcal{L}(x;\sigma,b)}$. In the case when $\sigma\sigma^*(.) > 0$, this inclusion does not provide any information. What we will show in the present section is that, in fact, if $\sigma\sigma^*(.) > 0$, then $\text{supp}\,(P_x) = \Omega_x \equiv \{w \in \Omega.: x(0,w) = 0\}$. Actually, we are going to derive a somewhat more general result.

(3.1) LEMMA: *Given* $\varepsilon > 0$, *define*

$$u_\varepsilon(t,x) = \sum_{n=-\infty}^{\infty} \int_{-\varepsilon}^{\varepsilon} [\gamma_t(x-y-4n\varepsilon) - \gamma_t(x-y-(4n+2)\varepsilon)]dy \tag{3.2}$$

for $t > 0$ *and* $x \in (-\varepsilon,\varepsilon)$, *where*

$$\gamma_t(z) = \frac{1}{(2\pi t)^{1/2}} e^{-z^2/2t}.$$

Then

$$u_\varepsilon \in C^\infty((0,\infty) \times R^1), \quad \frac{\partial u_\varepsilon}{\partial t} = \frac{1}{2} \frac{\partial^2 u_\varepsilon}{\partial x^2} \quad \textit{in } (0,\infty) \times R^1,$$

$\lim_{t \downarrow 0} u_\varepsilon(t,x) = 1$ *for* $x \in (-\varepsilon,\varepsilon)$, $u_\varepsilon(t,x) > 0$ *for* $(t,x) \in (0,\infty) \times (-\varepsilon,\varepsilon)$,

$u(t, \pm\varepsilon) = 0$ *for* $t > 0$ *and* $\frac{\partial^2 u_\varepsilon}{\partial x^2}(t,x) \leq 0$ *for* $(t,x) \in (0,\infty) \times (-\varepsilon,\varepsilon)$.

Finally, if $(\beta(t), F_t, P)$ *is a one dimensional Brownian motion,* $T > 0$ *and* $x \in (-\varepsilon,\varepsilon)$, *then*

(3.3) $$u_\varepsilon(T,x) = P(x + \beta(t) \in (-\varepsilon,\varepsilon) \text{ for } t \in [0,T]).$$

Proof: Let

$$S = \bigcup((4n-1)\varepsilon\ ,\ (4n+1)\varepsilon)$$

and

$$S' = 2\varepsilon + S = \{2\varepsilon + z : z \in S\}.$$

Then

$$u_\varepsilon(t,x) = \int_S \gamma_t(x-y)\,dy - \int_{S'} \gamma_t(x-y)\,dy\ .$$

This proves that

$$u_\varepsilon \in C^\infty((0,\infty) \times R^1)$$

and that

$$\frac{\partial u_\varepsilon}{\partial t} = \frac{1}{2}\frac{\partial^2 u_\varepsilon}{\partial x^2}$$

for $t > 0$. Also, if $x \in (-\varepsilon,\varepsilon)$, then $x \in S$ and $x \notin \overline{S}'$.

Hence

$$\lim_{t \downarrow 0} \int_S \gamma_t(x-y)\,dy = 1 \quad \text{and} \quad \lim_{t \downarrow 0} \int_{S'} \gamma_t(x-y)\,dy = 0.$$

Thus

$$\lim_{\varepsilon \downarrow 0} \int_S u_\varepsilon(t,x) = 1 \quad \text{for } x \in (-\varepsilon,\varepsilon)\ .$$

Next observe that

$$u_\varepsilon(t,\varepsilon) = \int_{S+\varepsilon} \gamma_t(y)\,dy - \int_{S+\varepsilon} \gamma_t(y)\,dy = 0.$$

Since $u_\varepsilon(t,.)$ is clearly even, this shows that $u_\varepsilon(t, \pm\varepsilon) = 0$ for $t > 0$.

We next prove (3.3). To this end, let

$$\tau_x = \inf \{t \geqslant 0 : x + \beta(t) \notin (-\varepsilon,\varepsilon)\} .$$

Then, by Itô's formula applied to $u_\varepsilon(T - t, x)$:

$$u_\varepsilon(T,x) = E[u_\varepsilon(T - \tau_x \wedge T, x + \beta(\tau_x \wedge T)$$

$$= P(\tau_x > T) = P(x+\beta(t) \in (-\varepsilon,\varepsilon) \quad \text{for} \quad t \in [0,T]).$$

From (3.3) it is clear that

$$\frac{\partial u_\varepsilon}{\partial t}(t,x) \leqslant 0,$$

and so

$$\frac{\partial^2 u_\varepsilon}{\partial x^2} = 2 \frac{\partial u_\varepsilon}{\partial t} \leqslant 0.$$

Finally, we must show that $u_\varepsilon(t,x) > 0$ for all $t > 0$ and $x \in (-\varepsilon,\varepsilon)$. To this end, suppose that $x^\circ \in (-\varepsilon,\varepsilon)$ and that $u_\varepsilon(T, x^\circ) = 0$ for some $T > 0$. Then we would have that

$$E[e^{z\tau_{x^\circ}}]$$

is an entire function of $z \in \mathbb{C}$. On the other hand for a given $0 < \lambda < \pi/2^{3/2}\varepsilon$, consider the function

$$\phi_\lambda(x) = \frac{\cos(2^{1/2}\lambda\, x)}{\cos(2^{1/2}\lambda\, \varepsilon)} .$$

Applying Itô's formula to $e^{\lambda^2 t}\,\phi_\lambda(x)$, one sees that

$$\phi_\lambda(x\ ) = E[e^{\lambda^2 \tau_{x^\circ}}]$$

and so

$$\lim_{\lambda \uparrow \pi/2^{3/2}\varepsilon} E[e^{\lambda^2 \tau_{x^\circ}}] = \infty\ ,$$

which clearly is a contradiction.

(3.4) LEMMA: *Let* (E, F, P) *be a probability space,* $\{F_t : t \geq 0\}$ *a non-decreasing family of sub* σ *- algebras of* F *and* $\eta : [0,\infty) \times E \to R$ *a* P- *a.s. continuous* $F.$ *-progressively measurable function. Assume that there exist bounded* $F.$ *- progressively measurable functions* $a: [0,\infty) \times E \to [0,\infty)$ *and* $b: [0,\infty) \times E \to R$ *such that*

$$(f(t,\eta(t)) - \int_0^t (\frac{\partial f}{\partial s} + \frac{1}{2}\, a(s)\, \frac{\partial^2 f}{\partial x^2} + b(s)\, \frac{\partial f}{\partial x})\ (s,\eta(s))ds,\ F_t,\ P)$$

is a martingale for all $f \in C_b^{1,2}([0,\infty) \times R^1)$. *Let* σ *be an* $F.$*-stopping time and given* $\varepsilon > 0$ *define*

$$\tau_\varepsilon = \inf\,\{t \geq \sigma : |\eta(t) - \eta(\sigma)| \geq \varepsilon\}\ .$$

If $b(s) \equiv 0$ *for* $\sigma \wedge T \leq s \leq T \wedge \tau_\varepsilon$ *and* $a(s) \leq A$ *for* $\sigma \wedge T \leq s \leq T \wedge \tau_\varepsilon$ *then*

$$P(\tau_\varepsilon \leq T) \leq (1 - u_\varepsilon(AT,\ 0))P(\sigma \leq T).$$

<u>Proof</u>: Let $F: [0,\infty) \times R^1 \times E \to R^1$ have the properties that for each $(t,x) \in [0,\infty) \times R^1$ $F(t,x)$ is F_σ - measurable and for each $q \in E$, $F(t,x,q) \in C_b^{1,2}([0,\infty) \times R^1$. Then by Doob's stopping time

theorem plus elementary properties of conditional expectations (cf 1.5.7 in [S & V]),

$$E[F(T \wedge \tau_\varepsilon, \eta(T \wedge \tau_\varepsilon)) - F(\sigma \wedge T, \eta(\sigma \wedge T)) | F_\sigma]$$
$$= E[\int_{\sigma \wedge T}^{\tau_\varepsilon \wedge T} (\frac{\partial F}{\partial s} + \frac{1}{2} a(s) \frac{\partial^2 F}{\partial x^2}) (s, \eta(s) | F_\sigma]$$

(a.s., P) on $\{\sigma < \infty\}$.

In particular, with $F(t,x) = u_\varepsilon(A(T - T \wedge t), x - \eta(\sigma))$ ($\equiv 0$ if $\sigma = \infty$), we have

$$P(\tau_\varepsilon > T, \sigma \leq T) = E(u_\varepsilon(A(T-T \wedge \tau_\varepsilon), \eta(\tau_\varepsilon \wedge T) - \eta(\sigma)), \sigma \leq T]$$
$$= E[F(T \wedge \tau_\varepsilon, \eta(T \wedge \tau_\varepsilon)), \sigma \leq T]$$
$$= E[F(\sigma \wedge T, \eta(\sigma \wedge T)), \sigma \leq T]$$
$$+ E[\int_{\sigma \wedge T}^{T \wedge \tau_\varepsilon} (\frac{\partial F}{\partial s} + \frac{1}{2} a(s) \frac{\partial^2 F}{\partial x}(s, \eta(s)) ds, \sigma \leq T].$$

Since $F(\sigma \wedge T, \eta(\sigma \wedge T)) = u_\varepsilon(A(T - \sigma), 0) \geq u_\varepsilon(AT, 0)$ on $\{\sigma \leq T\}$ and

$$(\frac{\partial F}{\partial s} + \frac{a(s)}{2} \frac{\partial F}{\partial x^2}) (s, \eta(s))$$
$$= (- A \frac{\partial u_\varepsilon}{\partial s} + \frac{a(s)}{2} \frac{\partial^2 u_\varepsilon}{\partial x^2}) (A(T-s), \eta(s) - \eta(\sigma))$$
$$= - \frac{1}{2} (A - a(s)) \frac{\partial^2 u_\varepsilon}{\partial x^2} (A(T-s), \eta(s) - \eta(\sigma)) \geq 0$$

for $\sigma \wedge T \leq s \leq T \wedge \tau_\varepsilon$, we see that

$$P(\tau_\varepsilon > T, \sigma \leq T) \geq u_\varepsilon(AT, 0) P(\sigma \leq T).$$

Hence

$$P(\tau_\varepsilon \leq T) = P(\sigma \leq T,\ \tau_\varepsilon \leq T)$$
$$= P(\sigma \leq T) - P(\sigma \leq T,\ \tau_\varepsilon > T)$$
$$< (1 - u_\varepsilon(AT, 0))\ P(\sigma \leq T).$$

(3.5) LEMMA: *Let* $(\beta(t), F_t, P)$ *be a d-dimensional Brownian motion,* $\alpha(.)$ *a bounded* $F.$ *- progressively measurable* $R^n \otimes R^d$ *- valued function,* $b(.)$ *a bounded* $F.$*- progressively measurable* R^n *- valued function and* $c(.)$ *a bounded* $F.$ *- progressively measurable* R^d*- valued function such that* $c(t) \equiv 0$ *for* $t \geq T$. *Define*

(3.6) $$R = \exp\left[\int_0^\infty < c(s),\ d\beta(s) > - \frac{1}{2}\int_0^\infty |c(s)|^2 ds\right].$$

Then $R > 0$ (a.s., P) $E^P[R] = 1$ *and if* $dQ = RdP$, *then for every* $f \in C_b^{1,2}([0,\infty), R^n)$

$$(f(t,\xi(t)) - \int_0^t (\frac{\partial f}{\partial s} + L_s f + < \alpha(s)c(s), \text{grad } f>)(s,\xi(s))ds,\ F_t,\ Q)$$

is a martingale, where

$$L_s \equiv \frac{1}{2}\sum_{i,j=1}^n (\alpha\alpha^*(s))^{ij} \frac{\partial^2}{\partial x_i \partial x_j} + \sum_{i=1}^n b^i(s)\frac{\partial}{\partial x_i}.$$

Proof: It is clear that $R > 0$ (a.s., P). Furthermore, by Itô's formula, if

$$R(t) = \exp\left[\int_0^t < c(s),\ d\beta(s)) > - \frac{1}{2}\int_0^t |c(s)|^2 ds\right],$$

then

(3.7) $$R(t) = 1 + \int_0^t R(s) < c(s),\ d\beta(s) >,\ t \geq 0.$$

Thus $E^P[R] = E^P[R(T)] = 1$. Finally, another application of Itô's formula shows that

$$\left(R(t)f(t,\xi(t)) - \int_0^t \left(R(s)\left(\frac{\partial f}{\partial s} + L_s f + \langle \alpha(s)c(s), \text{grad } f\rangle\right)(s,\xi(s))ds, F_t, P\right)$$

is a martingale. Since, from (3.7)

$$E^P[R|\ F_t] = R(t) \qquad \text{(a.s., } P)$$

for all $t \geq 0$, it follows immediately that

$$\left(f(t,\xi(t)) - \int_0^t \left(\frac{\partial f}{\partial s} + L_s f + \langle \alpha(s)c(s), \text{grad } f\rangle\right)(s,\xi(s))ds, F_t, Q\right)$$

is a martingale.

(3.8) THEOREM: *Let* $(\beta(t), F_t, P)$ *be a* d*-dimensional Brownian motion and let* $\alpha(.)$ *be a bounded* $F.$ *- progressively measurable* $R^n \otimes R^d$ *- valued function and* $\gamma(.)$ *a bounded* $F.$*- progressively measurable* R^d*- valued function. Set*

$$\xi(t) = \int_0^t \alpha(s)\, d\beta(s) + \int_0^t \alpha(s)\gamma(s)ds$$

and assume that

$$\frac{\text{Trace } \alpha\alpha^*(s)}{|\alpha\ (s)\ \ (s)|} \chi_{[\varepsilon,\infty)}(|\xi(s)|) \quad (\equiv 0 \quad \text{if} \quad \alpha(s) = 0)$$

is uniformly bounded for each $\varepsilon > 0$. *Then for each* $T > 0$ *and* $\varepsilon > 0$,

$$P(\sup_{0 \leq t \leq T} |\xi(t)| < \varepsilon) > 0.$$

Proof: Set $\eta(t) = |\xi(t)|^2$ and define

$$\sigma = \inf \{t \geq 0 : |\eta(t)| \geq 2\varepsilon\}$$

and

$$\tau_\varepsilon = \inf \{t \geq 0 : |\eta(t) - \eta(\sigma)| \geq \varepsilon\}.$$

Then

$$P(\sup_{0 \leq t \leq T} |\xi(t)| \geq (3\varepsilon)^{1/2}) = P(\sup_{0 \leq t \leq T} |\eta(t)| \geq 3\varepsilon)$$

$$\leq P(\tau_\varepsilon \leq T).$$

Thus we must show that $P(\tau_\varepsilon \leq T) < 1$. Without loss of generality, we will assume that $\alpha(S) \equiv 0$ if

$$|\xi(s)| \geq (3\varepsilon)^{1/2} .$$

To prove that $P(\tau_\varepsilon \leq T) < 1$, first note that

$$\eta(t) = 2 \int_0^t < \alpha^*(s)\xi(s), d\beta(s) > + \int_0^t \text{Trace } \sigma\sigma^*(s)ds + 2\int_0^t < \alpha^*(s)\xi(s), \gamma(s) > ds.$$

Now define

$$c(s) = -\gamma(s) - \frac{1}{2} \left(\frac{\text{Trace } \alpha\alpha^*(s)}{|\alpha^*(s)\xi(s)|^2} \alpha^*(s)\xi(s)\right)\chi_{[0,T]}(s)\chi_{[-\varepsilon/2,\infty]}(\eta(s)).$$

Then $c(.)$ is uniformly bounded and $c(s) \equiv 0$ for $s \geq T$. Set

$$R = \exp \left[\int_0^\infty < c(s), d\beta(s) > - \frac{1}{2} \int_0^\infty |c(s)|^2 ds\right]$$

and $dQ = RdP$. Since Q and P are mutually absolutely continuous

$$Q(\tau_\varepsilon \leq T) < 1 \text{ iff } P(\tau_\varepsilon \leq T) < 1.$$

But, by (3.5),

$$(f(t,\eta(t)) - \int_0^t (\frac{\partial f}{\partial s} + \frac{1}{2}a(s)\frac{\partial^2 f}{\partial x^2} + b(s)\frac{\partial f}{\partial x})(s,\eta(s)ds, F_t, Q)$$

is a martingale for all $f \in C_b^{1,2}([0,\infty) \times R^1)$, where

$$a(s) = 4\,|\alpha^*(s)\,\xi(s)|^2$$

and

$$b(s) = \text{Trace}\ \alpha\alpha^*(s) + 2 < \alpha^*\ \xi(s), c(s) > + <\alpha^*(s)\xi(s), \gamma(s)> .$$

In particular, $a(.)$ and $b(.)$ are bounded and $b(s) = 0$ for $\sigma \wedge T \leq s \leq \tau_\varepsilon \wedge T$. Thus, by (3.4), $Q(\tau_\varepsilon \leq T) < 1$.

(3.9) <u>COROLLARY</u>: *Let* $(\beta(t), F_t, P)$ *be a d-dimensional Brownian motion,* $\alpha(.)$ *a bounded* $F.$ *- progressively measurable* $R^n \otimes R^d$ *- valued function and* $v(.)$ *a bounded* $F.$ *- progressively measurable* R^n *-valued function. Assume that* $\alpha\alpha^*(.) \geq \lambda I$ *for some* $\lambda > 0$. *Set*

$$\xi(t) = x + \int_0^t \alpha(s)\,d\beta(s) + \int_0^t v(s)ds, \quad t \geq 0.$$

Then for every $\phi \in C_b^1[0,\infty), R^n)$ *satisfying* $\phi(0) = x$, *every* $\varepsilon > 0$ *and every* $T > 0$:

$$P(\sup_{0 \leq t \leq T} |\xi(t) - \phi(t)| < \varepsilon) > 0.$$

<u>Proof</u>: Given $\phi(.)$, set

$$\xi_\phi(t) = \xi(t) - \phi(t).$$

Then

$$\xi_\phi(t) = \int_0^t \alpha(s)d\ (s) + \int_0^t \alpha(s)\gamma(s)ds$$

where

$$\gamma(s) = \alpha^*(s)\ (\alpha\alpha^*(s))^{-1}(v(s) - \phi'(s)).$$

Note that

$$|\alpha^*(s)\xi(s)|^2 = \langle \xi(s), \alpha\alpha^*(s)\xi(s)\rangle \geq \lambda\, |\xi(s)|^2,$$

and so

$$\frac{\text{Trace } \alpha\alpha^*(s)}{|\alpha^*(s)\ \xi(s)|}\ \chi_{[\varepsilon,\infty)}(|\xi(s)|)$$

is uniformly bounded for each $\varepsilon > 0$.

Thus, by theorem (3.8),

$$P\Big(\sup_{0\leq t\leq T} |\xi_\phi(t)| < \varepsilon\Big) > 0$$

for each $\varepsilon > 0$ and $T > 0$.

(3.10) <u>COROLLARY</u>: *Let* $\sigma : R^n \to R^n \otimes R^d$ *and* $b: R^n \to R^n$ *be* C_b^∞ *- functions and set*

$$L = \frac{1}{2} \sum_{i,j=1}^{n} (\sigma\sigma^*)^{ij}(x) \frac{\partial^2}{\partial x_i \partial x_j} + \sum_{i=1}^{n} b^i(x) \frac{\partial}{\partial x_i} .$$

Assume that $\sigma\sigma^*(x) > 0$ *for each* $x \in R^n$. *Given,* $x \in R^n$, *let* $P_x \sim L$ *at* x. *Then*

$$\text{supp}\,(P_x) = \{\phi \in C([0,\infty), R^n) : \phi(0) = x\} .$$

<u>Proof</u>: As we already know, P_x is the distribution of the process $\xi(.,x)$ given by

$$\xi(t,x) = x + \int_0^t \sigma(\xi(s,x))d\beta(s) + \int_0^t b(\xi(s,x))ds$$

where $(\beta(t), F_t, P)$ is any d- dimensional Brownian motion. In particular , we may assume that

$$\beta(,) = (\bar{\beta}_1(.),\ldots, \bar{\beta}_d(.))$$

where $(\bar{\beta}(t), F_t, P)$ is a $(d+n)$ - dimensional Brownian motion. Now let $\phi \in C_b^1([0,\infty), R^n)$ with $\phi(0) = x$ be given. Set

$$K = \sup_{0 \leq t \leq T} |\phi(t)| + 2\varepsilon$$

and define

$$\bar{\xi}(t) = x + \int_0^t \bar{\sigma}(\xi(s,x))d\bar{\beta}(s) + \int_0^t b(\xi(s,x))ds, \; t \geq 0,$$

where $\bar{\sigma} : R^n \to R^n \otimes R^{d+n}$ is given by

$$\bar{\sigma}(y) = (\sigma(y), \chi_{[K,\infty)}(y)I) .$$

Then $\bar{\xi}(t) = \xi(t,x)$ for $0 \leq t \leq \tau_K \equiv \inf\{s \geq 0 : |\xi(s,x)| \leq K\}$. In particular,

$$P(\sup_{0<t<T} |\xi(t,x)-\phi(t)| < \varepsilon) = P(\sup_{0\leq t\leq T} |\xi(t,x)-\phi(t)| < \varepsilon, \tau_K > T)$$

$$= P(\sup_{0\leq t\leq T} |\bar{\xi}(t)-\phi(t)| < \varepsilon, \tau_K > T)$$

$$= P(\sup_{0\leq t\leq T} |\bar{\xi}(t)-\phi(t)| < \varepsilon).$$

But $\bar{\sigma}\bar{\sigma}^*(y) = \sigma\sigma^*(y) + \chi_{[K,\infty)}(y)I \geq \lambda I$ where

$$\lambda = \inf\{< \theta,\sigma\sigma^*(y)\theta > /|\theta|^2 : |y| \leq K \text{ and } \theta \in R^n \setminus \{0\}\} > 0.$$

Thus, by corollary (3.9),

$$P(\sup_{0\leq t\leq T} |\bar{\xi}(t) - \phi(t)| < \varepsilon) > 0.$$

We have therefore proved that

$$\{\phi \in C_b^1([0,\infty); R^n): \ \phi(0) = x\} \subseteq \operatorname{supp}(P_x) .$$

Since $\operatorname{supp}(P_x)$ is closed in Ω, this completes the proof.

(3.11) COROLLARY: (The Strong Maximum Principle): *Let* L *be as in* (3.10) *and let* G *be an open subset of* $R^1 \times R^n$. *Suppose that* $u \in C^{1,2}(G)$ *satisfies*

$$\frac{\partial u}{\partial t} + Lu \geq 0$$

in G *and that* $(t_0, x^\circ) \in G$ *has the property that* $u(t_0, x^\circ) \geq u(t,x)$ *for all* $(t,x) \in G$. *Then* $u(t_1, x^1) = u(t_0, x^\circ)$ *for all* $(t_1, x^1) \in G(t_0, x^\circ)$, *where* $G(t_0, x^\circ)$ *is the closure of the points* $(t_1, \phi(t_1 - t_0))$ *such that* $t_1 \geq t_0$, $\phi \in C([0,\infty); R^n)$, $\phi(0) = x^\circ$ *and* $(t, \phi(t - t_0)) \in G$ *for* $t_0 \leq t \leq t_1$. *In particular, if* U *is a connected open set in* R^n *and* $u \in C^2(U)$ *satisfies* $Lu \geq 0$ *in* U, *then* u *is constant in* U *if* u *attains its maximum in* U.

Proof: By replacing G by $\{(t-t_0, x) : (t,x) \in G\}$ and $u(t,x)$ by $u(t - t_0, x)$, we will assume that $t_0 = 0$. Furthermore, by approximating G from inside with relatively compact open regions, we will assume that $u \in C_b^{1,2}(G)$.

Note that by Itô's formula and Doob's stopping time theorem

$$E^{P_{x^\circ}}[u(t \wedge \tau, x(t \wedge \tau))] - u(0, x^\circ)$$
$$= E^{P_{x^\circ}}\left[\int_0^{t \wedge \tau} \left(\frac{\partial u}{\partial s} + Lu\right)(s, x(s))ds\right] \geq 0,$$

where $P_{x^\circ} \sim L$ at x° and $\tau = \inf\{t \geq 0 : (t, x(t)) \notin G\}$.

Thus

$$E^{P_{x^0}}[u(t \wedge \tau, x(t \wedge \tau)) - u(0, x^0)] \geq 0, \ t \geq 0.$$

Now suppose that $\phi \in C([0,\infty); R^n)$, $\phi(0) = x$ and $(t, \phi(t)) \in G$ for $0 \leq t \leq t_1$. If $u(t_1, \phi(t_1)) < u(0,x^0)$, then we could find an $\varepsilon > 0$ such that dist. $((s, \phi(s)), G^c) > \varepsilon$ for all $0 \leq s \leq t$ and $u(t_1, x) \leq u(0, x) - \varepsilon$ for $|x - \phi(t_1)| < \varepsilon$. Thus we would have

$$E^{P_{x^0}}[u(t_1 \wedge \tau, x(t_1 \wedge \tau)) - u(0, x^0)]$$

$$\leq - \varepsilon P_{x^0}(\sup_{0 \leq t \leq t_1} |x(t) - \phi(t)| < \varepsilon) < 0,$$

which is a contradiction.

(3.12) Remark: The preceding result can be extended in the following way:

Let $a : R^1 \times R^n \to R^n \otimes R^n$ and $b: R^1 \times R^n \to R^n$ be measurable functions such that a is symmetric and uniformly positive definite on compact sets and a and b are bounded on compact sets. Set

$$L_t = \frac{1}{2} \sum_{i,j=1}^{n} a^{ij}(t,x) \frac{\partial^2}{\partial x_i \partial x_j} + \sum_{i=1}^{n} b^i(t,x) \frac{\partial}{\partial x_i} .$$

If $u \in C^{1,2}(G)$ satisfies

$$\frac{\partial u}{\partial t} + Lu \geq 0$$

in G and u attains its maximum at $(t_0, x^0) \in G$,

then $u(t_1, x^1) = u(t_0, x^\circ)$ for all $(t_1, x^1) \in G(t_0, x)$. The proof can be constructed along the same lines as we have just used, the only missing ingredient is a more sophisticated treatment of the existence theory for solutions to the martingale problem (cf [Berk. Symp.] for the details).

§4. THE SUPPORT OF $P_x \sim L$, THE DEGENERATE CASE:

We are now going to show that if

$$(4.1) \qquad L = \frac{1}{2} \sum_{\ell=1}^{d} \left(\sum \sigma_\ell^i(x) \frac{\partial}{\partial x_i} \right)^2 + \sum_{i=1}^{n} b^i(x) \frac{\partial}{\partial x_i} \, ,$$

and if $P_x \sim L$ at x, then

$$(4.2) \qquad \operatorname{supp}(P_x) = \overline{\mathcal{S}(x;\, \sigma,\, b)}.$$

Since we already know that $\operatorname{supp}(P_x) \subseteq \overline{\mathcal{S}(x;\, \sigma,\, b)}$, it suffices to show that if

$$(4.3) \qquad \phi(t) = x + \int_0^t \sigma(\phi(s))\, \dot{\psi}(s)\,ds + \int_0^t b(\phi(s))\,ds, \quad t \geq 0,$$

where $\psi \in C_0^2((0,\infty);\, R^d)$, then

$$(4.4) \qquad P_x\left(\sup_{0 \leq t \leq T} |x(t) - \phi(t)| \leq \varepsilon \right) > 0$$

for all $\varepsilon > 0$ and $T > 0$. Actually, what we are going to prove is slightly more refined result than (4.4). Namely, suppose that $(\beta(t), F_t, P)$ is a d - dimensional Brownian motion and that

$$(4.5) \qquad \xi(t,x) = x + \int_0^t \sigma(\xi(s,x))\,d\beta(s) + \int_0^t \tilde{b}(\xi(s,x))\,ds, \quad t \geq 0,$$

where

(4.6) $$\tilde{b}(x) = b(x) + \frac{1}{2}\sum_{j=1}^{n}\sum_{\ell=1}^{d}\sigma_\ell^j(x)\frac{\partial\sigma_\ell}{\partial x_j}(x) .$$

Then P_x is the distribution of $\xi(.,x)$. We will show that

(4.7) $$\lim_{\delta\downarrow 0} P(\sup_{0\le t\le T}|\xi(t,x)-\phi(t)| \geq \varepsilon | \sup_{0\le t\le T}|\beta(t)-\psi(t)| \le \delta) = 1$$

for all $\varepsilon > 0$ and $T > 0$. Since

$$P(\sup_{0\le t\le T}|\beta(t) - \psi(t)| \le \delta) > 0$$

for $\delta > 0$ and $T > 0$, this will certainly prove (4.4). The proof of (4.7) relies on a few facts about Brownian motion.

(4.8) <u>LEMMA</u>: *Let* $(\beta(t), F_t, P)$ *be a* d *-dimensional Brownian motion. Then there exist* $A > 0$ *and* $B > 0$, *depending on* d , *such that*

(4.9) $$P(\sup_{0\le t\le T}|\beta(t)| < \delta) \geq A\exp(-\frac{BT}{\delta^2}),\ T > 0 \text{ and } \delta > 0.$$

<u>Proof</u>: First note that if

$$\Phi(T,\delta) = P(\sup_{0\le t\le T}|\beta(t)| < \delta),$$

then

$$\Phi(T,\delta) = \Phi(T/\delta^2, 1).$$

The reason for this is that for any $\lambda > 0$, $\lambda\beta(.)$ has the same distribution as $\beta(\lambda^2,.)$ (cf. exercise (4.11) below). Thus we need only check that

$$\Phi(t, 1) \geq A e^{-Bt}, t > 0.$$

Next observe that

$$P(\sup_{0\le t\le T}|\beta(t)| < \delta) \geq P(\sup_{0\le t\le T}\max_{1\le\ell\le d}|\beta^\ell(t)| < \delta/ d^{1/2})$$

$$= P(\sup_{0 \leq t \leq T} |\beta^1(t)| < \delta/d^{1/2})^d$$

since the $\beta^\ell(.)$'s are mutually independent and each has the same distribution. Thus we will restrict our attention to the case when $d = 1$.

To prove that $\Phi(T,1) \geq A\, e^{-BT}$ when $d = 1$, we will show that if $f \in C([-1, 1])$

$$\text{(4.10)} \quad \begin{aligned} &E[f(x + \beta(T)),\ \sup_{0 \leq t \leq T} |x + \beta(t)| < 1] \\ &= \sum_{m=0}^{\infty} a_m(f)\, e^{(-m^2\pi^2/8)T} \sin(m\pi/2)\,(x + 1)), \end{aligned}$$

$$T > 0 \text{ and } x \in (-1, 1),$$

where

$$a_m(f) = \int_0^1 f(y)\sin((m\pi/2)(y+1))dy / \int_0^1 \sin^2((m\pi/2)(y+1))dy.$$

Given (4.10), we will have

$$P(\sup_{0 \leq t \leq T} |\beta(t)| < 1) = a_1(1)\, e^{(-\pi^2/8)T} + \sum_{m=1}^{\infty} a_{2m+1}(1)(-1)^{m+1} e^{(-(2m+1)^2\pi^2/8)T}$$

Since $a_1(1) > 0$ and $|a_m(1)| \leq 1$ for all m, it is clear from this that

$$\lim_{T \uparrow \infty} e^{(\pi^2/8)T}\, P(\sup_{0 \leq t \leq T} |\beta(t)| < 1) = a_1(1) > 0,$$

and so our estimate will be established. To prove (4.10), first note that

$$\{\sin \tfrac{m\pi}{2}(x+1) : m \geq 1\}$$

is an orthonormal basis in $L^2((-1, 1))$ (cf. exercise (4.11) below).

Next observe that for, $f \in C_0^\infty((-1, 1))$,

$$\sum_{m=0}^{\infty} a_m(f) \sin (m\pi/2 \ (x+1))$$

is uniformly and absolutely convergent to f and that if $u(T,x)$ is given by the right hand side of (4.10) then

$$\frac{\partial u}{\partial T} = \frac{1}{2} \frac{\partial^2 u}{\partial x^2}, \quad T > 0 \quad \text{and} \quad x \in R^1,$$

$$\lim_{T \downarrow 0} u(T, x) = f, \ -1 < x < 1$$

$$u(T, \pm 1) = 0.$$

Hence, $(u(T - t \wedge T, x + \beta(t)), F_t, P)$ is a martingale and so

$$u(T, x) = E[f(x) + \beta(T)), \tau_x > T]$$

where $\tau_x = \inf \{t \geqslant 0 : |x + \beta(t)| \geqslant 1\}$. Thus (4.10) holds for $f \in C_0^\infty((-1, 1))$. Using obvious limit procedure, it is now easy to see that (4.10) continues to hold for all $f \in ([-1, 1])$ so long as $T > 0$.

(4.11) <u>Exercise</u>: Fill the missing details in the preceding proof. In particular, use the martingale problem characterization of Brownian motion to check that $\lambda\beta(\cdot/\lambda^2)$ has the same distribution as $\beta(\cdot)$ for any $\lambda \in R \setminus \{0\}$. Second, show that

$$\{\sin (\frac{m\pi}{2} (x+1)) : m \geqslant 1\}$$

is an orthogonal basis in $L^2((-1, 1))$, that

$$\sup_m |a_m(f)| \leq c\|f\|_{L^2((-1,1))} ,$$

and that

$$\sum_{m=1}^{\infty} a_m(f) \sin(\frac{m\pi}{2}(x+1))$$

is absolutely and uniformly convergent to f if $f \in C_0^\infty((-1,1))$. All these facts are easy consequences of the elementary theory of Fourier series.

(4.12) LEMMA: *Let* $(B(t), F_t, P)$ *be a* 1- *dimensional Brownian motion. Then* P - a.s., *there is precisely one solution to the equation*

$$\xi(t) = 2\int_0^t |\xi(s)|^{1/2} dB(s) + 2t, \ t \geq 0.$$

Moreover, the unique solution $\xi(.)$ *is non-negative and* $B(.)$ *-measurable.*

Proof: For $0 < \varepsilon \leq 1$ and $n \geq 1$, consider the equation

$$\xi_n(t,\varepsilon) = \varepsilon + 2\int_0^t \sigma_n(\xi_n(s,\varepsilon))dB(s) + 2t ,$$

where $\{\sigma_n\}_1^\infty \subseteq C^\infty(R^1)$, $\sup_{n\geq 1} \sup_{|x|\leq 1} |\sigma_n(x)| \leq 1$ and $\sigma_n(x) = |x|^{1/2}$ for $|x| \geq \varepsilon/n$. Since $\sigma_n(.)$ is uniformly Lipschitz continuous, $\xi_n(.,\varepsilon)$ is uniformly determined and $B(.)$ - measurable. Moreover, if

$$\tau_n = \inf\{t \geq 0 : \xi_n(t,\varepsilon) \leq \varepsilon/n\} ,$$

then $\xi_{n+1}(t,\varepsilon) = \xi_n(t,\varepsilon)$, $0 \leq t \leq \tau_n$ (a.s., P). Next note that $\tau_n \uparrow \infty$ (a.s., P) as $n \to \infty$. To see this, define

$$\zeta_{n,R} = \inf\{t \geq 0 : |\xi_n(t,\varepsilon)| \geq R\}$$

for $R > \varepsilon$. Then

$$\overline{\lim_{n \to \infty}} P(\tau_n \leq \zeta_{n,R}) = 0 \quad \text{for all} \quad R > 0.$$

Indeed, choose $u \in C_b^2(R^1)$ so that

$$u(x) = \log (nx/\varepsilon)^2 \ / \ \log (nR/\varepsilon)^2$$

for $\varepsilon/n \leq x \leq R$. Then, by Itô's formula

$$P(\tau_n > \zeta_{n,R}) = E[u(\xi_n(\tau_n \wedge \zeta_{n,R}))] = u(\varepsilon)$$

$$= \log (n^2) \ / \ \log (nR/\varepsilon)^2 \to 1$$

as $n \to \infty$. Next note that by Doob's inequality, for any $T > 0$

$$P(\sup_{0 \leq t \leq T} |\xi_n(t,\varepsilon) - \varepsilon - 2t| \geq R)$$

$$\leq \frac{1}{R} E[|\xi_n(T,\varepsilon) - \varepsilon - 2T|]$$

$$\leq \frac{c(T)}{R}$$

where $c(T)$ is independent of n. Hence

$$\sup_{n \geq 1} P(\zeta_{n,R} \leq T) \to 0$$

as $R \to \infty$, and so we conclude that

$$\overline{\lim} P(\tau_n \leq T) = 0$$

for all $T > 0$.

We have now proved that there is for each $\varepsilon > 0$, P - a.s. a unique continuous $\xi(.,\varepsilon)$ such that $\xi(t,\varepsilon) = \xi_n(t,\varepsilon)$, $0 \leq t \leq \tau_n$, and that $\tau_n \uparrow \infty$ (a.s., P) as $n \to \infty$. Clearly $\xi(.,\varepsilon)$ is $B(.)$ - measurable and non-negative (a.s., P). Also

$$\xi(t,\varepsilon) = \varepsilon + 2 \int |\xi(s,\varepsilon)| dB(s) + 2t, \; t \geq 0.$$

We next show that

$$P(\sup_{0 \leq t \leq T} |\xi(t,\varepsilon) - \xi(t, \varepsilon')| \geq \lambda) \to 0$$

as ε , $\varepsilon' \to 0$ for each $T > 0$ and $\lambda > 0$. To this end, note that

$$0 < \rho(\delta) = \sup_{\substack{x,y \leq 1 \\ |x-y| \leq \delta}} (|x|^{1/2} - |y|^{1/2}) \leq C(\delta^{1/2} \wedge \delta), \; \delta > 0.$$

Thus we can find $\{\alpha_k\} \subseteq (0,1)$ so that

$$\int_{\alpha_k}^{\alpha_{k-1}} \frac{1}{\rho^2(\lambda)} \, d\lambda = k .$$

Choose $\{\phi_k''\} \subseteq C_0^\infty((\alpha_k, \alpha_{k-1}))$ so that $0 \leq \phi_k''(\cdot) \leq \frac{2}{k\rho^2(\cdot)}$

and

$$\int_0^\infty \phi_k'' (\lambda) \, d\lambda = 1.$$

Set

$$\phi_k(\lambda) = \int_0^{|\lambda|} dt \int_0^t \phi_k''(s) \, ds.$$

Then, by Itô's formula:

$$E[\phi_k(\xi(T,\varepsilon) - \xi(T, \varepsilon'))]$$

$$\leq \phi_k(\varepsilon-\varepsilon') + \frac{1}{2} E[\int_0^T \rho^2(|\xi(t,\varepsilon)-\xi(t,\varepsilon')|)\phi_k''(\xi(t,\varepsilon)-\xi(t,\varepsilon'))dt]$$

$$\leq |\varepsilon - \varepsilon'| + T/k .$$

Since $\phi_k(x) \uparrow |x|$ as $k \uparrow \infty$, we now see that

$$E[|\xi(T,\varepsilon) - \xi(T, \varepsilon')|] \leq |\varepsilon - \varepsilon'|.$$

But $(\xi(t,\varepsilon) - \xi(t,\varepsilon'), F_t, P)$ is a martingale; and so, by Doob's inequality

$$P\Big(\sup_{0\leq t\leq T} |\xi(t,\varepsilon) - \xi(t,\varepsilon')| \geq \lambda\Big) \leq \frac{|\varepsilon-\varepsilon'|}{\lambda} \to 0.$$

We now see that there exist $\varepsilon_n \downarrow 0$ such that $\xi(\cdot,\varepsilon_n) \to \xi(\cdot)$ (a.s.,P) uniformly on finite intervals. Clearly $\xi(\cdot)$ is $B(\cdot)$ - measurable, P - a.s. non-negative and satisfies

$$\xi(t) = 2\int_0^t |\xi(s)|^{1/2}\, dB(s) + 2t,\ t \geq 0$$

(cf. Exercise (4.13) below). Finally, if $\eta(t)$ were a second process satisfying the same equation, then, using the functions ϕ_k constructed above, we would find that

$$\eta(T) = \xi(T) \quad \text{(a.s., P)}$$

for all $T > 0$.

(4.13) <u>Exercise</u>: Show that in fact $\xi(\cdot,\varepsilon) \leq \xi(\cdot,\varepsilon')$ (a.s., P) if $0 \leq \varepsilon < \varepsilon'$. The idea is to employ a variation on the ideas used to prove

$$P\Big(\sup_{0\leq t\leq T} |\xi(t,\varepsilon) - \xi(t,\varepsilon')| \geq \lambda\Big) \to 0 \text{ as } \varepsilon,\varepsilon' \to 0.$$

(4.14) <u>LEMMA</u>: *Let* $(\beta(t), F_t, P)$ *be a* d- *dimensional Brownian motion where* $d \geq 2$. *Given* $1 \leq i \neq j \leq d$, *define*

$$L_{ij}(t) = \int_0^t \frac{\beta_i}{(\beta_i^2 + \beta_j^2)^{1/2}}\, d\beta_j - \int_0^t \frac{\beta_j}{(\beta_i^2 + \beta_j^2)^{1/2}}\, d\beta_i$$

(here we take $\frac{\beta_i}{(\beta_i^2+\beta_j^2)^{1/2}} = \frac{\beta_j}{(\beta_i^2+\beta_j^2)^{1/2}} = 0$ if $\beta_i^2 + \beta_j^2 = 0$)

Then $(L_{ij}(t), F_t, P)$ *is a* 1 *- dimensional Brownian motion and the process* $L_{ij}(.)$ *is independent of the process* $\beta_i^2(.) + \beta_j^2(.)$.

In particular, if $\theta(.)$ *is a bounded* $F.$ *- progressively measurable* R *- valued process, then for* $M \geq 1$ *and* $\delta > 0$:

(4.15) $$P(\sup_{0\leq t\leq T} |\int_0^t \theta(u)|\beta(u)|dL_{ij}(u)| > M\delta | \sup_{0\leq t\leq T} |\beta(t)| < \delta)$$

$$\leq 2\exp\left(-\frac{M^2}{2T\|\theta\|_u^2}\right)$$

<u>Proof:</u> Without loss of generality, we will assume that $i = 1$ and $j = 2$. Let $L(t) = L_{12}(t)$. To see that $(L(t), F_t, P)$ is a 1-dimensional Brownian motion, note that by Itô's formula

$$(f(L(t)) - \int_0^t \tfrac{1}{2}f''(L(s))ds, F_t, P)$$

is a martingale for all $f \in C_b^2(R^1)$.

(Remember that $P(\int_0^\infty \chi_{\{0\}}(\beta_1^2(s) + \beta_2^2(s))ds > 0) = 0$

since
$$E[\int_\varepsilon^T \chi_{\{0\}}(\beta_1^2(s) + \beta_2^2(s))ds] = \int_\varepsilon^T dt \int_{\{0\}} \frac{1}{2\pi t} e^{-y^2/2t} dy = 0$$

for all $0 < \varepsilon < T < \infty$.)

We next show that $L(.)$ is independent of $\beta_1^2(.) + \beta_2^2(.)$.

Define
$$B(t) = \int_0^t \frac{\beta_1}{\beta_1^2+\beta_2^2} d\beta_1 + \int_0^t \frac{\beta_2}{\beta_1^2+\beta_2^2} d\beta_2$$

(with the same convention when $\beta_1^2 + \beta_2^2 = 0$).

Again use Itô's formula to show that

$$(f(L(t),\ B(t)) - \frac{1}{2}\int_0^t \Delta f(L(s),\ B(s)ds,\ F_t,\ P)$$

is a martingale for all $f \in C_b^2(R^2)$. Thus $((L(t),\ B(t)),\ F_t,\ P)$ is a 2-dimensional Brownian motion, and so $L(.)$ is independent of $B(.)$.

Finally, by Itô's formula, if $\xi(t) = \beta_1^2(t) + \beta_2^2(t)$ then

$$\xi(t) = 2\int_0^t (\beta_1(s)d\beta_1(s) + \beta_2(s)d\beta_2(s)) + 2t$$

$$= 2\int_0^t |\xi(s)|^{1/2}\ dB(s) + 2t$$

and so, by (4.12), $\xi(.)$ is $B(.)$ - measurable. Hence $L(.)$ is independent of $\beta_1^2(.) + \beta_2^2(.)$.

Finally, since $L(.)$ is independent of $\xi(.)$ and they both are independent of $\sum_3^d \beta_j^2(.)$, we can argue as in (I.4.3) to prove that

$$P(\sup_{0 \le t \le T} |\int_0^t \theta(s)|\beta(s)|dL(s)| \geq \lambda|\beta(.)|)$$

$$\leq 2\ \exp\ (-\frac{\lambda^2}{2\|\theta\|_u^2 \int_0^T |\beta(s)|^2})$$

where $\|\ \|_u$ denotes the uniform norm.

In particular

$$P\left(\sup_{0 \leq t \leq T} \left| \int_0^t \theta(s) |\beta(s)| dL(s) \right| \geq M\delta \quad \sup_{0 \leq t \leq T} |\beta(t)| \leq \delta \right)$$

$$\leq 2 \exp\left(- M^2/2T\|\theta\|_u^2\right)$$

(4.16) <u>Exercise</u>: Let $(\beta(t), F_t, P)$ be a 2-dimensional Brownian motion. Prove that for each $\varepsilon > 0$, $P(|\beta(t \vee \tau_\varepsilon)| > 0, t \geq 0) = 1$, where $\tau_\varepsilon = \inf\{t \geq 0: |\beta(t)| \geq \varepsilon\}$. Next show that $P(\tau_\varepsilon \downarrow 0$ as $\varepsilon \downarrow 0)=$ Conclude that $P(|\beta(t)| > 0$ for all $t > 0) = 1$. Thus P - a.s., $\theta(t) = \arg \beta(t)$ is well defined for all $t > 0$). In order to study $\theta(t)$, $t > 0$, write $z(t) = \beta_1(t) + i\beta_2(t)$. Given an analytic function f on $\mathbb{C}$, show that $df(z(t)) = f'(z(t))dz(t)$.

In particular, show that for fixed $t_0 > 0$

$$\log z(t) - \log z(t_0) = \int_{t_0}^t \frac{1}{|z(s)|} dB(s) + \int_{t_0}^t \frac{dL(s)}{|z(s)|}, \quad t \geq t_0,$$

where $B(.)$ and $L(.)$ are defined as in (4.14). Hence

$$\theta(t) - \theta(t_0) = \int_{t_0}^t \frac{dL(s)}{|z(s)|}, \quad t \geq 0.$$

The conditional distribution of $\theta(. \vee t_0) - \theta(t_0)$ given $|z(.)|$ is the same as the distribution of $\tilde{B}\left(\int_{t_0}^{.\vee t_0} |z(s)|^{-2} ds\right)$, where $\tilde{B}(.)$ is a 1-dimensional Brownian motion.

(4.17) Notation: If $\theta : [0,\infty) \times R^N$ and $T > 0$, we will use $\| \theta(.) \|_T^0$ to denote $\sup_{0 \leq t \leq T} |\theta(t)|$.

(4.18) LEMMA: *Let* $(\beta(t), F_t, P)$ *be a* d*-dimensional Brownian motion and suppose that* $\theta(.)$ *is an* P - a.s. *continuous* $F.$ - *progressively measurable* R^d - *valued function with the property that for some* $\alpha > 1$

$$\lim_{M \to \infty} \sup_{0 < \delta < 1} P(\| \theta(.) \|_T^0 \geq M\delta^\alpha \,|\, \| \beta(.) \|_T^0 \leq \delta) = 0.$$

Then

$$\lim_{\delta \downarrow 0} P(\| \int_0^{\cdot} < \theta(u), d\beta(u) > \|_T^0 > \varepsilon \,|\, \| \beta(.) \|_T^0 \leq \delta)$$

is 0 *for all* $\varepsilon > 0$.

Proof: Let $0 < \delta < 1$ and $M \geq 1$ be given and define

$$\zeta = \inf \{t \geq 0 : |\theta(t)| \geq M\delta^\alpha\}.$$

Then

$$P(\| \int_0^{\cdot} < \theta(u), d\beta(u) > \|_T^0 \geq \varepsilon, \| \beta(.) \|_T^0 \leq \delta)$$

$$\leq P(\| \int_0^{\cdot} < \theta(u), d\beta(u) > \|_T^0 \geq \varepsilon, \| \theta(.) \|_T^0 \leq M\delta^\alpha)$$

$$+ P(\| \theta(.) \|_T^0 \geq M\delta^\alpha, \| \beta(.) \|_T^0 \leq \delta)$$

$$\leq P(\| \int_0^{\cdot} < \theta(u \wedge \zeta), d\beta(u) > \|_T^0 \geq \varepsilon)$$

$$+ P(\| \theta(.) \|_T^0 \geq M\delta^\alpha, \| \beta(.) \|_T^0 \leq \delta).$$

By the argument used to prove (I.4.3) :

$$P(\| \int_0^{\cdot} < \theta(u \wedge \zeta), d\beta(u) > \|_T^0 \geq \varepsilon) \leq 2 \exp\left(- \frac{\varepsilon^2}{M^2 \delta^{2\alpha} T}\right) .$$

Thus, by (4.9):

$$P(\|\int_0^{\cdot} < \theta(u), d\beta(u) >\|_T^0 \geq \varepsilon \quad \|\beta(.)\|_T^0 \leq \delta)$$

$$\leq \frac{2}{A} \exp(-\frac{\varepsilon^2}{M^2\delta^2\alpha_T} + \frac{BT}{\delta^2}) + P(\|\theta(.)\|_T^0 \geq M\delta^{\alpha} \quad \|\beta(.)\|_T^0 < \delta)$$

By assumption, given $\lambda > 0$ we can choose $M_\lambda < \infty$ so that the second term is less than $\lambda/2$ for $0 < \delta < 1$. We can then choose $0 < \delta_\lambda < 1$ so that the first term is less than $\lambda/2$ for this choice of M_λ.

(4.19) <u>LEMMA:</u> *Let* $(\beta(t), F_t, P)$ *be a* d*-dimensional Brownian motion and let* $\xi(.,x)$ *be given by* (4.5). *Then for any* $\varepsilon > 0$, $T > 0$, $f \in C_b^\infty(R^n)$ *and* $1 \leq k \leq d$ *or* $1 \leq k \neq \ell \leq d$:

$$\lim_{\delta \downarrow 0} P(\|\int_0^{\cdot} f(\xi(u,x))d(\beta^k(u))^2\|_T^0 \geq \varepsilon \quad \|\beta(.)\|_T^0 < \delta) = 0$$

and

$$\lim_{\delta \downarrow 0} P(\|\int f(\xi(u,x)) \beta^k(u)d\beta^\ell(u)\|_T^0 \geq \varepsilon \quad \|\beta(.)\|_T^0 < \delta) = 0.$$

<u>Proof</u>: *Set* $\xi(.) = \xi(.,x)$. *Then for* $1 \leq k, \ell \leq d$, *we see, by Itô's formula, that*

$$\int_0^t f(\xi(u))d(\beta^k\beta^\ell(u)) = \beta^k\beta^\ell(t)f(\xi(t))$$

$$- \int_0^t \beta^k\beta^\ell(u) < \sigma^*(\xi(u)) \text{ grad } f(\xi(u)), d\beta(u) >$$

$$- \int_0^t \beta^k\beta^\ell(u) Lf(\xi(u))du$$

$$- \int_0^t [\beta^k(u)(\sigma^*(\xi(u))\text{grad } f(\xi(u))_\ell + \beta^\ell(u)(\sigma^*(\xi(u))\text{grad } f(\xi(u))_k]du.$$

All except the second term on the right clearly tend to zero as $\|\beta(\cdot)\|_T^0 \to 0$. Moreover, the second term is covered by (4.18) with $\alpha = 2$. Thus

$$\lim_{\delta \downarrow 0} P(\|\int_0^{\cdot} f(\xi(u))d(\beta^k\beta^\ell(u))\|_T^0 \geq \varepsilon \mid \|\beta(\cdot)\|_T^0 < \delta) = 0$$

for all $\varepsilon > 0$ and $1 \leq k, \ell \leq d$. This proves our first assertion upon taking $k = \ell$.

To prove the second assertion, note that

$$\int_0^t f(\xi(u))\beta^k(u)d\beta^\ell(u) = \frac{1}{2}\int f(\xi(u))d(\beta^k(u)\beta^\ell(u)) + \frac{1}{2}\int_0^t \theta(u)|\beta(u)|dL_{k,\ell}(u) ,$$

where $L_{k,\ell}(\cdot)$ is as in (4.14) and

$$\theta(\cdot) \equiv \frac{(\beta^k(\cdot)^2 + \beta^\ell(\cdot)^2)^{1/2}}{|\beta(\cdot)|} f(\xi(u)).$$

Thus the first term is of the sort just treated and the second one is covered by (4.14).

(4.20) THEOREM: *Let* $(\beta(t), F_t, P)$ *be a d-dimensional Brownian motion and let* $\xi(\cdot,x)$ *be given by* (4.5). *Given* $\psi \in C_0^2((0,\infty); R^d)$, *define* $\phi(\cdot)$ *by* (4.3). *Then for each* $\varepsilon > 0$ *and* $T > 0$:

$$\lim_{\delta \downarrow 0} P(\sup_{0\leq t\leq T} |\xi(t,x)-\phi(t)| \geq \varepsilon \mid \sup_{0\leq t\leq T} \|\beta(t)-\psi(t)\| \leq \delta)=1.$$

In particular, if $P_x \sim L$ *at* x, *where* L *is defined by* (4.1), *then*

$$\mathrm{supp}\,(P_x) = \overline{\mathcal{S}(x;\, \sigma, b)} .$$

Proof: We first note that it suffices to handle the case when $\psi \equiv 0$ but $b(.)$ may depend on t as well as x. Indeed, the general case reduces to this one by considering the probability measure Q defined by $dQ = R_\psi \, dP$, where

$$R_\psi = \exp\left(\int_0^\infty < \dot{\psi}(s), d\beta(s) > - \frac{1}{2}\int_0^\infty |\dot{\psi}(s)|^2 \, ds\right).$$

By Lemma (3.5), $(\beta_\psi(t), F_t, Q)$ is a d-dimensional Brownian motion, where $\beta_\psi(t) = \beta(t) - \psi(t)$, $t \geq 0$ and clearly

$$\xi(t,x) = x + \int_0^t \sigma(\xi(s,x))d\beta_\psi(s) + \int_0^t \tilde{b}_\psi(s,\xi(s,x)), \; t \geq 0,$$

(a.s., Q), where $\tilde{b}_\psi(t,x) = \tilde{b}(x) + \sigma(x)\dot{\psi}(t)$. Thus, assuming the case when $\psi \equiv 0$, we have

$$\lim_{\delta \downarrow 0} Q\left(\|\xi(.,x) - \phi(.)\|_T^0 \geq \varepsilon \;\; \|\beta(.) - \psi(.)\|_T^0 < \delta\right) = 1$$

and so

$$\overline{\lim_{\delta \downarrow 0}} \; P(\|\xi(.,x) - \phi(.)\|_T^0 \geq \varepsilon \;\; \|\beta(.) - \psi(.)\|_T^0 < \delta)$$

$$= \lim_{\delta \downarrow 0} \frac{E^P[\chi_{[\varepsilon,\infty)}(\|\xi(.,x)-\phi(.)\|_T^0)\chi_{[0,\delta]}(\|\beta(.)-\psi(.)\|_T^0)]}{E^P[R_\psi(T)\chi_{[\varepsilon,\infty)}(\|\xi(.,x)-\phi(.)\|_T^0)\chi_{[0,\delta]}\;\|\beta(.)-\psi(.)\|_T^0)]} \times$$

$$\times \frac{E^P[R_\psi(T)\chi_{[0,\delta)}(\|\beta(.) - \psi(.)\|_T^0)]}{E^P[\chi_{[0,\delta]}(\|\beta(.) - \psi(.)\|_T^0)]}$$

where

$$R_\psi(T) = \exp\left(\int_0^T < \dot\psi(s), d\beta(s) > - \frac{1}{2}\int_0^T |\dot\psi(s)|^2\, ds\right]$$

$$= \exp\left(< \psi(T), \beta(T) > - \int_0^T < \beta(s), \ddot\psi(s) > ds - \frac{1}{2}\int_0^T |\dot\psi(s)|^2 ds\right).$$

Observe that the ratio in the first factor tends to

$$\exp\left(- |\dot\psi(T)|^2 + \int_0^t < \psi(s), \ddot\psi(s) > ds + \frac{1}{2}\int_0^T |\dot\psi(s)|^2 ds\right) \quad \text{as} \quad \delta \downarrow 0$$

while the ratio in the second factor tends to

$$\exp\left(|\dot\psi(T)|^2 - \int_0^t < \psi(s), \ddot\psi(s) > ds - \frac{1}{2}\int_0^T |\dot\psi(s)|^2 ds\right. ;$$

and so the products tends to 1.

To handle the case when $\psi \equiv 0$ and b depends on (t,x), let $\xi(t) = \xi(t,x)$. Then

$$\xi^i(t) = x^i + \sigma^i_\ell(\xi(t))\beta^\ell(t) - \int_0^t \beta^\ell(s)\sigma^i_{\ell,j}(\xi(s))\sigma^j_k(\xi(s))d\beta^k(s)$$
$$- \int_0^t \beta^\ell(s)(L\sigma^i_\ell)(\xi(s))ds - \frac{1}{2}\int_0^t \sigma^i_{\ell,j}(\xi(t))\sigma^i_\ell(\xi(s))ds$$
$$+ \int_0^t b^i(s,\xi(s))ds$$
$$= x^i + \int_0^t b^i(s,\xi(s))ds - \Delta^i(t) ,$$

where

$$\Delta^i(t) = \sigma^i_\ell(\xi(t))\beta^\ell(t) - \sum_{k\neq\ell}\int_0^t \sigma^i_{\ell,j}\sigma^j_k(\xi(s))\beta^\ell(s)d\beta^k(s)$$
$$- \frac{1}{2}\int_0^t \sigma^i_{k,j}\sigma^i_k(\xi(s))d(\beta^k(s))^2 - \int_0^t \beta^\ell(s)(L\sigma^i)(\xi(s))ds.$$

We have used in these expressions the convention that repeated indices are summed and the notation

$$\sigma^i_{k,j} = \frac{\partial \sigma^i_k}{\partial x_j} .$$

By Lemma (4.19),

$$P(\|\Delta(.)\|^0_T \geq \varepsilon \quad \|\beta(.)\|^0_T \leq \delta) \to 0 \quad \text{as} \quad \delta \downarrow 0$$

for each $\varepsilon > 0$ and $T > 0$. Hence

$$\lim_{\delta \downarrow 0} P(\|\xi(.) - x - \int_0^{\cdot} b(s,\xi(s))ds\|^0_T \geq \varepsilon \quad \|\beta(.)\|^0_T \leq \delta) = 0$$

But, since $|b(t,x) - b(t,y)| \leq c|x-y|$, it is easily seen from this that

$$\lim_{\delta \downarrow 0} P(\|\xi(.) - \phi(.)\|^0_T \geq \varepsilon \quad \|\beta(.)\|^0_T \leq \delta) = 0,$$

where

$$\phi(t) = x + \int_0^t b(s,\phi(s))ds, \quad t \geq 0.$$

(4.21) <u>COROLLARY</u>: *Let* G *be an open set in* $R \times R^n$ *and suppose that* $u \in C^{1,2}(G)$ *satisfies*

$$\frac{\partial u}{\partial t} + Lu \geq 0$$

in G, *where* L *is given by* (4.1). *If* $(t_0, x^o) \in G$ *and* $u(t_0, x^o) = \max_G u$, *then* $u(t,x) = u(t_0, x^o)$ *at all points* $(t,x) \in G_L(t_0, x^o)$, *where* $G_L(t_0, x^o)$ *is the closure in* G *of the points* $(t_1, \phi(t_1 - t_0))$, $t_1 \geq t_0$, *where* $\phi \in \mathcal{S}(x^o; \sigma, b)$ *satisfies* $(t, \phi(t - t_0)) \in G$ *for* $t_0 \leq t \leq t_1$. *In fact, if*

$$u(t_0, x) = \max_{G_L(t_0,x^0)} u ,$$

then $u \equiv u(t_0, x)$ *on* $G_L(t_0, x^0)$.

Proof: Given (4.20), the proof is precisely the same as that of (3.11).

(4.22) Remark: When $G = R \times R^n$, it is easy to show that if $(t_1, x^1) \notin G_L(t_0,x^0)$ then there is a $u \in C^{1,2}(G)$ whose maximum is achieved at (t_0, x^0) and yet $u(t_1,x^1) < u(t_0, x^0)$. If $t_1 < t_0$, this is easy: simply take

$$u(t,x) = \begin{cases} -\exp\left(-\frac{1}{t_0-t}\right) & \text{if } t < t_0 \\ 0 & \text{if } t \geq t_0 \end{cases}$$

If $t_1 \geq t_0$, choose an open set $U \ni (t_1, x^1)$ so that $U \cap G_L(t_1, x^1) = \phi$ and let $f \in C_0^\infty(U)$ be a non-positive function such that $f(t_1, x^1) < 0$. Set

$$u(t, x) = E\left[\int_0^\infty f(s, \xi(s-t, x))ds\right] .$$

Using (I, 4.6), one can easily show that $u \in C_b(R \times R^n)$ and that

$$\frac{\partial u}{\partial t} + Lu = -f \geq 0.$$

Moreover, by (4.20), $u(t_0, x^0) = 0$. Finally, $u(t_1, x^1) < 0$.

In order to show that $G_L(t_0, x^0)$ is maximal when $G \neq R \times R^n$, one must extend the notion of "$\partial u/\partial t + Lu \geq 0$" to functions u which are not necessarily smooth. This is done in the paper [Degen. Diff.'s]. In that same paper, it is also shown how to extend the characterization of supp (P_x) to $P_x \sim L$ at x when L cannot be written in the form (4.1).

§5. THE "MOST PROBABLE PATH" OF A BROWNIAN MOTION WITH DRIFT:

Let $(\beta(t), F_t, P)$ be a d-dimensional Brownian motion and let $b: R^d \to R^d$ be a C^∞ - vector field with bounded first derivatives. Define $\xi(.)$ by

$$(5.1) \qquad \xi(t) = \beta(t) + \int_0^t b(\xi(s))\, ds, \quad t \geq 0.$$

We already know that for any $\phi \in C^2([0,\infty), R^d)$ with $\phi(0) = 0$,

$$P\left(\sup_{0 \leq t \leq T} \|\xi(t) - \phi(t)\| < \varepsilon \right) > 0$$

for all $\varepsilon > 0$ and $T > 0$. We now want to get an asymptotic estimate on this probability as $\varepsilon \downarrow 0$.

(5.2) LEMMA: *There is an orthonormal real basis*

$$\{\phi_n\}_0^\infty \subseteq C_b^\infty(B(0,1)) \quad \text{of} \quad L^2(B(0,1))$$

such that

$$\lim_{|x| \uparrow 1} \phi_n(x) = 0$$

and

$$-\frac{1}{2}\Delta\phi_n = \lambda_n \phi_n, \quad n \geq 0$$

where $0 < \lambda_0 < \lambda_1 \leq \lambda_2 < \cdots \leq \lambda_n \leq \cdots \uparrow \infty$. *Furthermore,* ϕ_0 *never vanishes in* $B(0,1)$. *Finally, there is an* $N = N(d)$ *such that*

$$\sum_0^\infty \frac{1}{\lambda_n^N} \phi_n(x)\, \phi_n(y)$$

converges absolutely and uniformly in $\overline{B(0,1)} \times \overline{B(0,1)}$; *and if,* $f \in C_b(B(0,1))$ *then*

(5.3) $$E[f(x+\beta(t)),\ \tau_x > t] = \sum e^{-\lambda_n t} (f,\phi_n)_{L^2(B(0,1))}\ \phi_n(x)\ ,$$

where

$$\tau_x = \inf\{t \geq 1 : |x+\beta(t)| \geq 1\}.$$

In particular,

$$P(\sup_{0 \leq t \leq T} |\beta(t)| < \varepsilon) \sim C\, e^{-\lambda T/\varepsilon^2} \text{ as } T/\varepsilon^2 \to \infty\ ,$$

where $C = (\phi_0, 1)_{L^2(B(0,1))}\ \phi_0(0)$ *and* $\lambda = \lambda_0$.

<u>Proof</u>: The spectral properties of $-\frac{1}{2}\Delta$ with Dirichlet boundary conditions in $(B(0,1))$ are well-known. In particular, the facts that the spectrum is completely discrete and positive and that λ_0 is simple can be found in elementary books on Partial Differential Equations. The absolute and uniform convergence of

$$\sum_0^\infty \frac{1}{\lambda_n^N}\ \phi_n(x)\ \phi_n(y)$$

for some N is a consequence of Mercer's theorem applied to the N-th iterate of the Green's function. From this it is easy to check that if $u(t,x)$ is given by the right hand side of (5.3) with $f \in C_0^\infty(B(0,1))$ then

$$\frac{\partial u}{\partial t} = \frac{1}{2}\Delta u \quad \text{in} \quad (0,\infty) \times B(0,1),$$

$$\lim_{|x| \uparrow 1} u(t,x) = 0 \quad \text{for} \quad t \geq 0, \quad \text{and}$$

$$\lim_{t \downarrow 0} u(t,x) = f(x), \quad x \in B(0,1).$$

Hence by Itô's formula, (5.3) holds for $f \in C_0^\infty(B(0,1))$. The general case is then proved by approximation.

Once one has (5.3), it is not hard to show that ϕ_0 never vanishes. Indeed, from (5.3), it is clear that

$$0 \leq e^{\lambda_0 t} P(\tau_x > t) = (\phi_0, 1)\, \phi_0(x) + 0(1) \quad \text{as} \quad t \uparrow \infty.$$

Thus if $(\phi_0, 1) > 0 (< 0)$, then $\phi_0 \geq 0 (\leq 0)$. From $\frac{1}{2} \Delta\phi_0 = -\lambda_0\phi_0$ and the strong maximum principle, it follows that $\phi_0 > 0 (< 0)$. On the other hand, if $(\phi_0, 1) = 0$, then from (5.3) with $f = \|\phi_0\|_u + \phi_0$:

$$0 \leq e^{\lambda_0 t} E[f(x + \beta(t)), \tau_x > t] = \phi_0(x) + O(1) ,$$

which obviously contradicts $(\phi_0, 1) = 0$.

Finally, as we saw in (4.8),

$$P(\sup_{0 \leq t \leq T} |\beta(t)| < \varepsilon) = P(\tau_0 > T/\varepsilon^2)$$

and clearly the above considerations prove that

$$P(\tau_0 > t) \sim C\, e^{-\lambda_0 t}$$

with

$$C = (\phi_0, 1)\, \phi_0(0) > 0.$$

(5.4) <u>Exercise</u>: Let G be a bounded, connected open set in R^d. Given $x \in G$, let $\tau_x = \inf \{t \geq 0: (x + \beta(t)) \notin G\}$. Show that

$$P(\tau_x > t) \sim C(x)\, e^{-\lambda t} \quad \text{as} \quad t \uparrow \infty ,$$

where $C(x) > 0$ and $\lambda > 0$ does not depend on x.

Now let $\phi \in C_0^2([0,\infty), R^d)$ and $T > 0$ be given and define

(5.5) $$\psi(t) = \phi(t) - \int_0^t b(\phi(s))ds, \quad t \geq 0$$

(5.6) $$b(t,x) = b(x + \phi(t)) - b(\phi(t)), \quad (t,x) \in [0,\infty) \times R^d ,$$

(5.7) $$\xi_\phi(t) = \xi(t) - \phi(t), \quad t \geq 0,$$

(5.8) $$\beta_\psi(t) = \beta(t) - \psi(t \wedge T), \quad t \geq 0$$

and

(5.9) $$R_\psi = \exp \Big[\int_0^T < \psi(s), d\beta(s) > - \frac{1}{2} \int_0^T |\psi(s)|^2 ds\Big].$$

Observe that

(5.10) $$\xi_\phi(t) = \beta_\psi(t) + \int_0^t b(s, \phi(s))ds, \quad t \geq 0 ;$$

and that if $dQ_\psi = R_\psi dP$, then $(\beta_\psi(t), F_t, Q_\psi)$ is a d-dimensional Brownian motion (cf. Lemma 3.5). Also

$$R_\psi = \exp [< \dot\psi(T), \beta(T) > - \int_0^T < \ddot\psi(t), \beta(t) > dt - \frac{1}{2} \int_0^T |\dot\psi(t)|^2 dt]$$

$$= \exp [< \dot\psi(t), (T) > - \int_0^T < \ddot\psi(t), \psi(t) > dt - \frac{1}{2} \int |\dot\psi(t)|^2 dt]$$

$$\times \exp[< \dot\psi(T), \beta_\psi(T) > - \int_0^T < \ddot\psi(t), \beta_\psi(t) > dt]$$

$$\to \exp [\frac{1}{2} \int_0^T |\dot\psi(t)|^2 dt]$$

uniformly and boundedly as $\|\beta_\psi(\cdot)\|_T \to 0$.

Since

$$\|\beta_\psi(\cdot)\|_T^0 < (1 + LT)\| \xi_\phi(\cdot)\|_T^0 ,$$

where $L = \|\text{grad } b\|_u$, we now have

$$\lim_{\varepsilon \downarrow 0} \frac{P(\|\xi_\phi(.)\|_T^O < \varepsilon)}{Q_\psi(\|\xi_\phi(.)\|_T^O < \varepsilon)} = \exp\left[-\frac{1}{2}\int_0^T |\dot\phi(t) - b(\phi(t))|^2 dt\right]$$

We now turn to the study of $Q_\psi(\|\xi_\phi(.)\|_T^O < \varepsilon)$.

(5.12) LEMMA: *Referring to the notation in* (5.5) - (5.9), *we have*

$$Q_\psi(\|\xi_\phi(.)\|_T^O < \varepsilon) = E[R(T) \;\|\; \beta(.)\|_T^O < \varepsilon] ,$$

where

$$R(T) = \exp\left[\int_0^T < b(s, \beta(s)), d\beta(s) > - \frac{1}{2}\int_0^T |b(s,\beta(s))|^2 ds\right]$$

Proof: First note that without loss of generality we may assume that $b \in C([0,\infty) \times R^n)$, since if this is not the case we can replace $b(s,x)$ by such a function b' which coincides with b on $\overline{[0;T]\times B(0,\|\phi(.)\|_T^O+\varepsilon)}$. Next observe that since $(\beta_\psi(.), F_t, Q_\psi)$ is a Brownian motion and $\xi_\phi(.)$ is given by (5.10), the distribution of $\xi_\phi(.)$ under Q_ψ is the same as the distribution of $\eta(.)$ under P, where

$$\eta(t) = \beta(t) + \int_0^t b(s, \eta(s))ds, \; t \geq 0.$$

But, if $dQ = R(T)\,dP$, then $\eta(.\wedge T)$ has the same distribution under P as $\beta(.\wedge T)$ has under Q. Hence

$$Q_\psi(\|\xi_\psi(.)\|_T^O < \varepsilon) = P(\|\eta(.)\|_T^O < \varepsilon)$$
$$= E^P[R(T) \;\|\beta(.)\|_T < \varepsilon]. \qquad \text{Q.E.D.}$$

Since we already know the asymptotics of $P(\|\beta(.)\|_T < \varepsilon)$ as $\varepsilon \downarrow 0$, it remains only to compute

$$\lim_{\varepsilon \downarrow 0} E^P[R(T) \;\|\beta(.)\|_T < \varepsilon] .$$

To this end, note that

$$E^{P}[R(T), \| \beta(.)\|_T^0 < \varepsilon] = e^{0(\varepsilon)} E^{P} [\exp[\int_0^T < b(s,\beta(s)), d\beta(s)> \ \| \beta\ (.)\|_T^0 < \varepsilon]]$$

Next

$$\int_0^T < b(s,\beta(s)),\ d\beta(s) > \ = \ <\beta(T),\ b(T,\ \beta(T)) > - \int_0^T (\frac{\partial}{\partial s} + \frac{1}{2}\Delta) b(s,\beta(s))ds$$

$$- \sum_{k,\ell} \int_0^T \beta_k(s)\ b^k_{,\ell}(\beta(s) + \phi(s))d\beta^{\ell}(s)$$

where

$$b^k_{,\ell}(x) \equiv \frac{\partial b^k}{\partial x_\ell}(x) \ .$$

Finally,

$$\sum_{\ell} \int_0^T \beta_k(s)\ b^k_{,\ell}(\beta(s) + \phi(s))d\beta^{\ell}(s)$$

$$= \frac{1}{2} \int_0^T \operatorname{div}\ b(\beta(s) + \phi(s))ds - \frac{1}{2} \sum_k \int_0^T b^k_{,k}(\beta(s)+\phi(s))d(\beta^k(s))^2$$

$$- \sum_{k \neq \ell} \int_0^T b^k_{,\ell}(\beta(s) + \phi(s))\ \beta^k(s)\ d\beta^{\ell}(s).$$

Combining all these, we arrive at

$$E^{P}[R(T) \ \ \| \beta(.)\|_T^0 < \varepsilon]$$

$$= e^{O(\varepsilon)} \exp(\frac{1}{2} \int_0^T \operatorname{div}\ b(\phi(s))ds) \times E^{P} \exp(\sum_k \Delta_k(T) + \sum_{k \neq \ell} \Delta_{k,\ell}(T)) \ \|\beta(.)\|_T^0 < \varepsilon]$$

where

$$\Delta_k(T) = -\frac{1}{2} \int_0^T b^k_{,k}(\beta(s) + \phi(s))d(\beta^k(s))^2$$

and

$$\Delta_{k,\ell}(T) = \int_0^T b^k_{,\ell}(\beta(s)+\phi(s))\beta^k(s)d\beta^\ell(s).$$

To complete our programme, we must prove that

$$\lim_{\varepsilon \downarrow 0} E^P[\exp(\sum_k \Delta_k(T) + \sum_{k \neq \ell} \Delta_{k,\ell}(T)) \quad \|\beta(.)\|_T^0 < \varepsilon) = 1.$$

(5.13) <u>LEMMA</u>: *Let* $f \in C_b^\infty([0,\infty) \times R^d; R^1)$. *Then for* $1 \leq k \leq d$ *or* $1 \leq k \neq \ell \leq d$

$$\sup_{0 \leq \varepsilon \leq 1} E^P[\exp(\int_0^T f(u,\beta(u))d(\beta^k(u))^2) \quad \|\beta(.)\|_T^0 < \varepsilon] < \infty$$

and

$$\sup_{0 \leq \varepsilon \leq 1} E^P[\exp(\int_0^T f(u,\beta(u))\beta^k(u)d\beta^\ell(u)) \quad \|\beta(.)\|_T^0 < \varepsilon] < \infty.$$

<u>Proof</u>: Let $1 \leq k, \ell \leq d$. Then

$$\int_0^T f(u,\beta(u))d(\beta^k\beta^\ell(u)) = \beta^k\beta^\ell(T)f(T,\beta(T))$$
$$- \int_0^T \beta^k\beta^\ell(T) < \mathrm{grad}_x f(s,\beta(s)), d\beta(s)>$$
$$- \int_0^T [\beta^k\beta^\ell(s)\left(\frac{\partial}{\partial s} + \frac{1}{2}\Delta\right)f + \beta^k(s)\frac{\partial f}{\partial x_k} + \beta^\ell(s)\frac{\partial f}{\partial x_\ell}](s,\beta(s))ds.$$

Thus for $0 \leq \varepsilon \leq 1$:

$$E^P[\exp(\int_0^T f(s,\beta(s))d(\beta^k(u)\beta^\ell(u))) \quad \|\beta(.)\|_T^0 < \varepsilon]$$
$$\leq CE^P[\exp(-\int_0^T \beta^k\beta^\ell(s) < \mathrm{grad}_x f(s,\beta(s)), d\beta(s)>) \quad \|\beta(.)\|_T^0 < \varepsilon].$$

But, as in the proof of (I.4.3)

$$P(-\int_0^T \beta^k\beta^\ell(s) < \mathrm{grad}_x\, f(s,\beta(s)), d\beta(s) > \geq R \,\Big|\, \|\beta(.)\|_T^0 < \varepsilon)$$

$$\leq \exp(-\frac{R^2}{2T\varepsilon^4 \|\mathrm{grad}_x\, f\|_u^2}), \; R > 0.$$

Thus for any $1 \leq k, \; \ell \leq d$:

$$\sup_{0 \leq \varepsilon \leq 1} E^P[\exp(\int_0^T f(s,\beta(s))d(\beta^k\beta^\ell(s))) \quad \|\beta(.)\|_T^0 < \varepsilon] ,$$

since

$$P(\|\beta(.)\|_T^0 < \varepsilon) \geq A\, e^{-BT/\varepsilon^2} .$$

This clearly proves the first assertion.

If $1 \leq k \neq \ell \leq d$, then

$$\int_0^T f(s,\xi(s))\beta^k(s)d\beta^\ell(s) = \frac{1}{2}\int_0^T f(s,\xi(s))d(\beta^k\beta^\ell(s))$$

$$+ \frac{1}{2}\int_0^T \theta(s)|\beta(s)|dL_{k,\ell}(s) ,$$

where $L_{k,\ell}(.)$ is as in (4.14) and

$$\theta(.) = \frac{(\beta^k(.)^2 + \beta^\ell(.)^2)^{1/2}}{|\beta(.)|} f(., \xi(.)).$$

By what we have just seen

$$\sup_{0 \leq \varepsilon \leq 1} E^P[\exp(\lambda \int_0^T f(s,\xi(s))d(\beta^k\beta^\ell(s))) \quad \|\beta(.)\|_T^0 < \varepsilon] < \infty$$

for any $\lambda > 0$. On the other hand, by (4.15);

$$P(\|\int_0^{\cdot} \theta(s)|\beta(s)|dL_{k,\ell}(s)\|_T^0 \geq R \quad \|\beta(.)\|_T^0 < \varepsilon)$$

$$< 2\exp\left(-\frac{R^2}{2T\|\theta\|_u^2\varepsilon^2}\right)$$

Thus

$$\sup_{0 < \varepsilon \leq 1} E\,[\exp(\lambda \int_0^T \theta(s)|\beta(s)|dL_{k,\ell}(s)) \quad \|\beta(.)\|_T^O < \varepsilon] < \infty$$

for any $\lambda > 0$. Clearly this completes the proof.

(5.14) THEOREM: *Let* $(\beta(t), F_t, P)$ *be a* d*-dimensional Brownian motion and suppose that* $\xi(.)$ *is given by* (5.1), *where* $b: R^d \to R^d$ *is* C^∞ *and has bounded first derivatives. Then for* $\phi \in C^2([0,\infty), R^d)$ *satisfying* $\phi(0) = 0$:

$$\lim_{\varepsilon \downarrow 0} P(\|\xi(.)-\phi(.)\|_T < \varepsilon) \sim C_0 e^{-\lambda_0 T/\varepsilon^2} \exp(\frac{1}{2}\int_0^T |\phi(t)-b(\phi(t))|^2 dt$$

$$+ \frac{1}{2}\int_0^T \operatorname{div} b(\phi(t))dt)$$

as $\varepsilon \uparrow 0$, *where* C_0 *and* λ_0 *are the numbers described in* (5.2).

Proof: We have seen that

$$P(\|\xi(.) - \phi(.)\|_T^O < \varepsilon \;\; \|\beta(.)\|_T^O < \varepsilon)$$

$$= e^{\mathcal{O}(\varepsilon)} \exp(-\frac{1}{2}\int_0^T |\dot{\phi}(t)-b(\phi(t))|^2 + \frac{1}{2}\int_0^T \operatorname{div} b(\phi(t))dt) \times$$

$$\times E^P \exp(\sum_k \Delta_k(T) + \sum_{k \neq \ell} \Delta_{k,\ell}(T)) | \; \|\beta(.)\|_T^O < \varepsilon] .$$

By (4.19) (with $x = 0$ and $\xi(t,x) = \beta(t)$, we know that

$$P(|\Delta_k(T)| \geq \alpha | \; \|\beta(.)\|_T^O < \varepsilon) \quad \text{and} \quad P(|\Delta_{k,\ell}(T)| \geq \alpha \;\; \|\beta(.)\|_T^O < \varepsilon)$$

tend to 0 as $\varepsilon \downarrow 0$ for each $\alpha > 0$. By (5.13), we know that for any $\lambda > 0$.

$$\overline{\lim_{\varepsilon \downarrow 0}} \; E^{P}[\exp [\lambda(\sum_{k}\Delta_{k}(T) + \sum_{k \neq \ell} \Delta_{k,\ell}(T))] \;\; \|\beta(\cdot)\|_{T}^{0} < \varepsilon]$$

is finite. Hence

$$\lim_{\varepsilon \downarrow 0} \; E^{P}[\exp [\sum_{k} \Delta_{k}(T) + \sum_{k \neq \ell} \Delta_{k,\ell}(T) \;\; \|\beta(\cdot)\|_{T}^{0} < \varepsilon] = 1.$$

Therefore

$$\lim_{\varepsilon \downarrow 0} \; P(\|\xi(\cdot) - \phi(\cdot)\|_{T}^{0} < \varepsilon \;\; \|\beta(\cdot)\|_{T}^{0} < \varepsilon)$$

$$= \exp (- \frac{1}{2} \int_{0}^{T} |\dot{\phi}(t) - b(\phi(t))|^{2} dt + \frac{1}{2} \int_{0}^{T} \operatorname{div} b(\phi(t)) dt).$$

The desired result therefore follows from (5.2).

(5.15) Remark: The result in (5.14) makes it possible to discuss the notion of the "most probable path" of the process $\xi(\cdot)$. Indeed, from (5.14), one sees that the most likely route followed by $\xi(\cdot)$ in going from $0 \to x \in R^{d}$ in time T is the path $\phi_0(\cdot)$ which minimizes

$$\frac{1}{2} \int_{0}^{T} |\dot{\phi}(t) - b(\phi(t))|^{2} dt = \frac{1}{2} \int_{0}^{T} \operatorname{div} b(\phi(t))\, dt$$

subjected to $\phi(0) = 0$ and $\phi(T) = x$. Using the usual techniques from the calculus of variations, it is easy to develop the Euler equation for this problem and thereby get an idea about what $\phi_0(\cdot)$ looks like. In particular, when $b \equiv 0$, it is clear that

$$\phi_0(t) = \frac{t}{T} x, \; 0 \leq t \leq T.$$

Next suppose that d = 1. By the calculus of variations :

$$\ddot{\phi}_0(t) - (b'.b) \circ \phi_0(t) + \frac{1}{2}b'' \circ \phi_0(t) = 0.$$

In general, this equation of course cannot be solved explicitly. However, after multiplying through by $\dot{\phi}_0$, one sees that

$$\dot{\phi}_0^2(t) - b^2 \circ \phi_0(t) + b \circ \phi_0(t) = \text{constant}.$$

Thus one can get some idea how $\phi_0(.)$ looks by using the "phase plane method". That is, if $p = \dot{\phi}_0(t)$ and $q = \phi_0(t)$, then

$$p^2 - b^2(q) + b'(q) = \text{constant}.$$

In the special case when $b(x) = \alpha x$, one can solve for $\phi_0(.)$ explicitly:

$$\ddot{\phi}_0 - \alpha^2 \phi_0 = 0$$

and so

$$\phi_0(t) = x \frac{\sinh \alpha t}{\sinh \alpha T}$$

In this case

$$P(\|\xi(.) - \phi_0(.)\|_T^0 < \varepsilon) \sim C\, e^{-\lambda T/\varepsilon^2} \exp(- \frac{\alpha x^2}{2} \frac{e^{-2\alpha T}}{1-e^{-2\alpha T}} + \frac{\alpha}{2} T) .$$

<u>B I B L I O G R A P H Y</u>

[S & V] : STROOCK, D.W. and S.R.S. VARADHAN, "*Multidimensional Diffusion Processes*", Springer-Verlag, Grundlehren # 233. (1979)

[Berk, Symp.] : STROOCK, D.W. and S.R.S. VARADHAN., "*On the support of Diffusion Processes, with applications to the Strong maximum principle*". Proc. 6th Berkeley Symp. on Math. Stat. and Prob., Vol.III (1970), pp.333-360.

[Deg. Diff's] : STROOCK, D.W. and S.R.S. VARADHAN., "*On degenerate elliptic-parabolic operators of second order and their associated diffusions*", Comm. Pure Appl. Math. XXV (1972), pp.651-714.

TATA INSTITUTE OF FUNDAMENTAL RESEARCH

LECTURES ON MATHEMATICS AND PHYSICS

General Editors : K.G. RAMANATHAN and B.V. SREEKANTAN

MATHEMATICS

1. On the Riemann Zeta-function. By K. Chandrasekharan*
2. On Analytic Number Theory. By H. Rademacher*
3. On Siegel's Modular Functions. By H. Maass*
4. On Complex Analytic Manifolds. By L. Schwartz*
5. On the Algebraic Theory of Fields. By K.G. Ramanathan*
6. On Sheaf Theory. By C.H. Dowker*
7. On Quadratic Forms. By C.L. Siegel*
8. On Semi-group Theory and its application to Cauchy's problem in Partial Differential Equations. By K. Yosida*
9. On Modular Correspondences. By M. Eichler*
10. On Elliptic Partial Differential Equations. By J.L. Lions*
11. On Mixed Problems in Partial Differential Equations and Representations of Semi-groups. By L. Schwartz*
12. On Measure Theory and Probability. By H.R. Pitt*
13. On the Theory of Functions of Several Complex Variables. By B. Malgrange*
14. On Lie groups and Representations of Locally Compact Groups. By F. Bruhat*
15. On Mean Periodic Functions. By J.P. Kahane*
16. On Approximation by Polynomials. By J.C. Burkill*
17. On Meromorphic Functions. By W.K. Hayman*

---

* Not available.

18. On the Theory of Algebraic Functions of One Variable. By M. Deuring*

19. On Potential Theory. By M. Brelot*

20. On Fibre Bundles and Differential Geometry. By J.L. Koszul*

21. On Topics in the Theory of Infinite Groups. By B.H. Neumann*

22. On Topics in Mean Periodic Functions and the Two-Radius Theorem. By J. Delsarte*

23. On Advanced Analytic Number Theory. By C.L. Siegel*

24. On Stochastic Processes. By K. Ito*

25. On Exterior Differential Systems. By M. Kuranishi*

26. On Some Fixed Point Theorems of Functional Analysis. By F.F. Bonsall*

27. On Some Aspects of p-adic Analysis. By F. Bruhat*

28. On Riemann Matrices. By C.L. Siegel*

29. On Modular Functions of one complex variable. By H. Maass

30. On Unique Factorization Domains. By P. Samuel*

31. On The Fourteenth Problem of Hilbert. By M. Nagata

32. On Groups of Transformations. By J.L. Koszul

33. On Geodesics in Riemannian Geometry. By M. Berger*

34. On Topics in Analysis. By Raghavan Narasimhan*

35. On Cauchy Problem. By Sigeru Mizohata

36. On Old and New Results on Algebraic Curves. By P. Samuel*

37. On Minimal Models and Birational Transformations of Two Dimensional Schemes. By I.R. Shafarevich

38. On Stratification of Complex Analytic sets. By M.-H. Schwartz

39. On Levi Convexity of Complex Manifolds and Cohomology Vanishing Theorems. By E. Vesentini

40. On An Introduction to Grothendieck's Theory of the Fundamental Group. By J.P. Murre

41. On Topics in Algebraic K-Theory. By Hyman Bass*

42. On the Singularities of the three-body problem. By C.L. Siegel

43. On Polyhedral Topology. By J.R. Stallings*

44. On Introduction to Algebraic Topology. By G. de Rham

45. On Quadratic Jordan Algebras. By N. Jacobson*

46. On The Theorem of Browder and Novikov and Siebenmann's Thesis. By M.A. Kervaire

47. On Galois Cohomology of Classical Groups. By M. Kneser*

48. On Discrete subgroups of Lie Groups. By G.D. Mostow

49. On the Finite Element Method. By Ph. Ciarlet

50. On Disintegration of measures. By L. Schwartz

51. On Introduction to moduli problems and orbit spaces. By P.E. Newstead.

52. On Numerical methods for time dependent equations - Application to Fluid flow problems. By P. Lascaux

53. On Optimisation - Theory and Algorithms. By Jean Cea

54. On Deformations of Singularities. By M. Artin

55. On Sieve Methods. By Hans E. Richert

56. On Irregularities of Distribution. By Wolfgang M. Schmidt

57. On Expansion techniques in Algebraic Geometry. By S.S. Abhyankar

58. On Torus Embeddings and Applications. By T. Oda

59. On Forms of Higher degree. By Jun. -Ichi Igusa

60. On Curves on rational and unirational surfaces. By M. Miyanishi

61. On Wave Propagation. By G.B. Whitham

62. On Equations defining space curves. By L. Szpiro

63. On Topics in Finite element solution of Elliptic Problems. By Bertrand Mercier

64. On Diffusion Problems and partial differential equations. By S.R.S. Varadhan

65. On Numerical Methods for Non-linear Variational Problems.
By R. Glowinski

66. On Representations of Complex Semi-simple Lie Groups.
By Thomas J. Enright

67. On Nonlinear Waves and Shocks. By Cathleen S. Morawetz

68. On Topics in Stochastic Differential Equations.
By Daniel W. Stroock

Errata

Page	Line (a = from above b = from below)	For	Read
2	9b	$(X(T)^{*})$	$(X(T)^{*})^{2}$
3	1a	$< C_p \cdots$	$\leq C_p$
3	6b	a(t)	a(t)dt
3	3b	a(t)	a(t)dt)
5	2a	$b(\xi_n(t)) - b(\xi_{n-1}(t)$	$(b(\xi_n(t)) - b(\xi_{n-1}(t)))$
5	7a	p-2/2p	(p-2)/2p
7	8b	$\int [\cdots]$	$\int \psi[\cdots]dy$
9	3b	$\leq$	$\geq$
10	4a,6a	B(O,R)	B(O,r)
11	5b	$u_n \leq v_n$	$u_n < v_n$
18	2b	$\dfrac{\cdots}{(1 + \frac{\lvert x\rvert}{\lvert x\rvert})(1 + \frac{\lvert y\rvert}{\lvert y\rvert})}$	$\dfrac{\cdots}{(1 + \frac{\lvert x\rvert}{\lvert x\rvert^2})(1 + \frac{\lvert y\rvert}{\lvert y\rvert^2})}$
18	2b	$\left\lvert\frac{y}{x}\right\rvert^2$	$\frac{\lvert y\rvert^2}{\lvert x\rvert^2}$
18	1b	$\left\lvert\frac{x}{y}\right\rvert$	$\frac{\lvert x\rvert}{\lvert y\rvert}$
19	1a	$\dfrac{\left\lvert\frac{x}{y}\right\rvert^2(\left\lvert\frac{y}{x}\right\rvert^2 -1)+\left\lvert\frac{x}{y}\right\rvert \lvert x-y\rvert}{\cdots}$	$\dfrac{\frac{\lvert x\rvert^2}{\lvert y\rvert^2}(\frac{\lvert y\rvert^2}{\lvert x\rvert^2} -1)+\frac{\lvert x\rvert}{\lvert y\rvert} \lvert x-y\rvert}{\cdots}$
22	7a	$\int_0^t \frac{\partial\hat{b}}{\partial\gamma_i} \cdots$	$\int_0^1 \frac{\partial\hat{b}}{\partial\gamma_i} \cdots$
22	8a	$\int_0^t \frac{\partial\hat{b}}{\partial\eta_j} \cdots$	$\int_0^1 \frac{\partial\hat{b}}{\partial\eta_j} \cdots$
22	4b	$\int_0^1 \frac{\partial\sigma}{\partial\gamma_i} \cdots$	$\int_0^1 \frac{\partial\hat{\sigma}}{\partial\gamma_o} \cdots$
25	1b	$S_k^2(\xi(s,x))J^{-1}(s,x)$	$S_k^2(\xi(s,x))$
27	1a	R^{n-1}	R^{n}
27	3b	Here $\Vert\ \Vert_u$ denotes the uniform norm	
32	2b	+ 0(1/n)	$+o(1/N)$.
36	11a	$\tau_{N(t)}/\tau_{N+1}$	$\tau_{N(t)} \mid \tau_{N(t)+1}$

Errata

Page	Line	For	Read
38	6a	$\frac{\varepsilon^2}{6} \phi''' \cdots$	$\frac{\varepsilon^3}{6} \phi''' \cdots$
39	4a	$(\Omega,\ )$	(Ω, M)
39	10a	$\overline{\lim}$	$\overline{\lim}_{\varepsilon \downarrow 0}$
41	6a,7a	S^{n-1}	S^{d-1}
41	9a,6b	R^n	R^N
41	2b	$h(A_\varepsilon(S),\theta(S))$	$(h(A_\varepsilon(s),\theta(S))$
42	6a	$M_\varepsilon(t)$	$M_{\varepsilon,\phi}(t)$
42	8a	S^{n-1}	S^{d-1}
42	8a	$h(A_\varepsilon(S),\eta)$	$\varepsilon h(A_\varepsilon(s),\eta)$
46	1a,7a	grad ϕ	$\text{grad}_x \phi$
46	4b	$x + \varepsilon \int_0^t \cdots$	$x + (\frac{d}{2})^{1/2} \varepsilon \int_0^t \cdots$
52	2b	F_σ	F_t
53	4a	$(s, \eta(s)$	$(s, \eta(s))ds$
53	6b	$\frac{\partial F}{\partial x^2}$	$\frac{\partial^2 F}{\partial x^2}$
57	5a	$\alpha^* \xi(s)$	$\alpha^*(s)\ \xi(s)$
65	9a	$f(x)$	$f(x$
65	8b	$([-1, 1])$	$C([-1, 1])$
68	1a	$\|\xi(s,\varepsilon)\|$	$\|\xi(s,\varepsilon)\|^{1/2}$
70	8b	> 0	$= 0$
77	6b	$\sigma^i_{\ell,j}(\xi(t))\sigma^i_\ell(\xi(s))$	$\sigma^i_{\ell,j}(\xi(t))\sigma^j_\ell(\xi(s))$
82	3a	$O(1)$	$o(1)$
82	7a	$O(1)$	$o(1)$
83	6a	$\psi(s)$	$\dot{\psi}(s)$
83	8a	$b(s, \phi(s))$	$b(s, \xi_\phi(s))$
86	6b	$\beta^k\beta^\ell(T)$	$\beta^k\beta^\ell(s)$
87	4a	$],$	$] < \infty,$